Animal Signals and Communication

Volume 1

Series Editors

Vincent M. Janik
School of Biology
University of St Andrews
Fife, UK

Peter Mc Gregor
Centre for Applied Zoology
Cornwall College
Newquay, UK

For further volumes:
http://www.springer.com/series/8824

Berthold Hedwig
Editor

Insect Hearing and Acoustic Communication

Springer

Editor
Berthold Hedwig
Department of Zoology
University of Cambridge
Cambridge
UK

ISSN 2197-7305 ISSN 2197-7313 (electronic)
ISBN 978-3-642-40461-0 ISBN 978-3-642-40462-7 (eBook)
DOI 10.1007/978-3-642-40462-7
Springer Heidelberg New York Dordrecht London

Library of Congress Control Number: 2013951135

Printed on acid-free paper

Springer is part of Springer Science+Business Media (www.springer.com)

Preface

This book provides a comprehensive selection of recent studies addressing the evolution, ecology, neurobiology and genetic analysis of hearing and acoustic communication in insects. Across a variety of hearing species, leading researchers in the field address the key points of current research.

The idea for this book emerged when the editor gave an Ernst Florey Memorial lecture at the 2011 Meeting of the German Neuroscience Society in Göttingen. I am most grateful to all my colleagues whose chapters are published here, who have contributed their time and effort as well as their chapters, to the Springer Team Mrs. A. König and Mrs. A. Schlitzberger for technical support and to the Series Editor P. McGregor for professional advice.

I would also like to thank my colleagues, at the Department of Zoology at the University of Cambridge and elsewhere, for their help and fruitful discussions. I thank my family for their support during the long hours necessary to develop this volume.

Finally, I would like to dedicate this book to Prof. Norbert Elsner who always fostered behavioral and neurobiological studies in insect bioacoustics at the University of Göttingen, and I hope the book will promote the field of bioacoustics among future generations of research students and scientists.

Berthold Hedwig

Preface

[illegible]

Contents

Chapter 1
Introduction

Berthold Hedwig

Abstract Research on insect hearing and acoustic communication has made enormous progress in the twentieth century. Following the first descriptions of auditory organs behavioural studies pointed to the importance of insect hearing for intraspecific acoustic communication, predator avoidance and prey detection. Analysing the neural mechanisms underlying auditory processing and the motor activity for acoustic signalling in a variety of species has provided us with a deep functional understanding of this insect behaviour. As questions of central biological importance are exemplified in *Insect Hearing and Acoustic Communication* these will drive the current and future research.

The sounds of singing and chirping insects are a salient feature of summer meadows, Mediterranean olive groves or tropical rain forests and have always caught the attention of naturalists and scientists. The evolution of insect ears has been driven by sexual and by natural selection in the context of intraspecific communication and by predator avoidance and prey detection. Ears have evolved in a variety of body regions, and evolved many times in parallel. The functional significance of insect acoustic signals and hearing organs, however, has only become evident over time.

When Pumphrey (1940) wrote the first comprehensive review on "Hearing in Insects", he began his conclusions with "The present time is perhaps a happy one for reviewing the experimental findings on the physiology of audition in insects in relation to their behaviour". For Pumphrey, due to the earlier first detailed anatomical descriptions of hearing organs in different groups of insects, for example by Schwabe (1906) and Eggers (1911), and the demonstration that female crickets are attracted to the acoustic signals of a male calling song broadcast by a telephone receiver (Regen 1913), it became obvious that sound production and the sense of hearing played a crucial role in insect behaviour.

B. Hedwig (✉)
Department of Zoology, University of Cambridge, Cambridge, UK
e-mail: bh202@cam.ac.uk

B. Hedwig (ed.), *Insect Hearing and Acoustic Communication*,
Animal Signals and Communication 1, DOI: 10.1007/978-3-642-40462-7_1,

The development of suitable electronic equipment allowed the field to progress rapidly, as indicated by the following key papers: The first electrophysiological recordings of auditory nerve activity in crickets and bushcrickets were obtained by Weever and Bray (1933) and gave an indication of the frequency range of the auditory responses. The biophysics of tympanal organs was analysed by Autrum (1941) and was related to a general theory of insect hearing. Roeder took his moth preparations to the field and demonstrated both the ultrasound sensitivity of moth ears and the afferent responses to sonar calls of hunting bats passing by (Roeder and Treat 1957). In 1961, the first recordings of thoracic auditory neurons in Tettigoniids were reported by Suga and Katsuki (1961) addressing frequency analysis, directional sensitivity and central inhibition. Questions about the higher central mechanisms of acoustic communication remained open for some time until Huber (1960) elicited singing behaviour in crickets by means of electrical brain stimulation and Roeder (1969) carried out the first recordings of auditory brain neurons in moths.

Then developments were fast: The advent of neural tracing techniques and intracellular recordings in insects allowed such advances as a systematic study of auditory pathways at the level of identified neurons, their structure and response properties (e.g. Rehbein et al. 1974; Casaday and Hoy 1977; Wohlers and Huber 1978) and a cellular approach to the neuronal control of singing behaviour (Bentley 1969). The increase in the inventory of scientific tools over the last decades with advances in behavioural and neural recording methods, in microscopy, and in molecular and genetic methods now allows tackling fundamental problems of insect hearing and acoustic communication at all levels from ecology to molecular mechanisms.

1.1 Central Research Questions

The variety of signalling behaviours and hearing organs makes insects highly suitable animals to explore and analyse signal generation and auditory processing. The fascinating progress that has been made is still related to a set of central questions characterising the focus of past, current and future research:

- How did hearing and acoustic communication behaviour evolve in insects and what is the neural and developmental origin of the auditory organs?
- What are the functional properties of hearing organs in respect of intensity, frequency and directional sensitivity and how are these achieved at a molecular, biophysical and neural level?
- How are hearing and sound production embedded in the natural lifestyle of the animals allowing intraspecific communication and also predator avoidance and even predation?
- How is phonotactic behaviour tuned to the communication signals of conspecifics? What are the neural mechanisms of peripheral and central auditory

processing that allow the recognition of species-specific sounds and lead to adapted motor responses?
- What are the biophysical and neural mechanisms underlying signal generation? How are central pattern generators organised that drive the species-specific motor activity and how is their activity controlled by the brain?
- Finally: What is the genetic basis of acoustic communication behaviour that leads to species-specific signal generation and pattern recognition and even speciation?

The subsequent chapters of this book will cover and address these questions to different degrees. The answers to these questions provide us with profound and fundamental understanding of a conspicuous and crucial insect behaviour. The final story will emerge by our ongoing research activity, closing again with Pumphrey (1940): "It will be obvious that much remains to be done".

References

Autrum H (1941) Über Gehör und Erschütterungssinn bei Locustiden. Zeitschr vergl Physiol 28(5):580–637

Bentley DR (1969) Intracellular activity in cricket neurons during the generation of behaviour patterns. J Ins Physiol 15:677–684

Casaday GB, Hoy RR (1977) Auditory interneurons in the cricket *Teleogryllus oceanicus*: physiological and anatomical properties. J Comp Physiol A 121(1):1–13

Eggers F (1911) Über das thoracale tympanale Organ der Noctuiden. Sitzungsberichte der Naturforscher-Gesellschaft bei der Universität Tartu (Dorpat) 30:138–145

Huber F (1960) Untersuchungen über die Funktion des Zentralnervensystems und insbesondere des Gehirnes bei der Fortbewegung und der Lauterzeugung der Grillen. J Comp Physiol A 44:60–132

Pumphrey R (1940) Hearing in insects. Biol Rev 15(1):107–132

Regen J (1913) Ueber die Anlockung des Weibchens von *Gryllus campestris* L. durch telephonisch übertragene Stridulationslaute des Männchens. Pflüg Arch f Physiol 155(1):193–200

Rehbein HG, Kalmring K, Römer H (1974) Structure and function of acoustic neurons in the thoracic ventral nerve cord of *Locusta migratoria* (Acrididae). J Comp Physiol A 95:263–280

Roeder KD, Treat EA (1957) Ultrasonic reception by the tympanic organ of noctuid moths. J Exp Zool 134:127–158

Roeder KD (1969) Acoustic interneurons in the brain of noctuid moths. J Ins Physiol 15(5):825–838

Schwabe J (1906) Beiträge zur Morphologie und Histologie der tympanalen Sinnesapparate der Orthopteren. Zoologica (Stuttgart) 20. Heft 50:1–154

Suga N, Katsuki Y (1961) Central mechanism of hearing in insects. J Exp Biol 38(3):545–558

Wever EG, Bray CW (1933) A new method for the study of hearing in insects. J Cell Comp Physiol 4(1):79–93

Wohlers DW, Huber F (1978) Intracellular recording and staining of cricket auditory interneurons (*Gryllus campestris* L., *Gryllus bimaculatus* DeGeer). J Comp Physiol A 127(1):11–28

Chapter 2
Evolutionary and Phylogenetic Origins of Tympanal Hearing Organs in Insects

Johannes Strauß and Reinhard Lakes-Harlan

Abstract Among insects, tympanal ears evolved at least 18 times, resulting in a diversity of auditory systems. Insects use their ears in different behavioural contexts, mainly intraspecific communication for mate attraction, predator avoidance, and parasitic host localisation. Analysing the evolution of insect ears aims at revealing the phyletic origins of auditory organs, the selection pressures leading to the evolution of ears, the physiological and behavioural adaptations of hearing, and the diversification of ears in specific groups or lineages. The origin of sensory organs from preadapted proprioceptive or vibroceptive organs has now been established for different ear types. In this review, we embed research on insect hearing in a phylogenetic framework to reconstruct the ancestral sensory situation in different taxa, and the series of morphological changes during the evolution of an ear. The importance of sensory and neuroanatomical data is discussed for either mapping onto a phylogeny or as characters for phylogenetic analysis.

2.1 Evolutionary Diversity of Insect Ears in Structure and Function

Insects are among the oldest land animals, and exist for more than 400 million years. For a large portion of this time, the majority of animal sounds stemmed from insects. Insects were probably also the first animals to evolve sound perceiving organs. Ears are found in different recent insect taxa and hearing is involved in

J. Strauß (✉) · R. Lakes-Harlan
AG Integrative Sensory Physiology, Institute for Animal Physiology,
Justus-Liebig-Universität Gießen, Heinrich-Buff-Ring 26, Gießen 35392, Germany
e-mail: johannes.strauss@physzool.bio.uni-giessen.de

R. Lakes-Harlan
e-mail: reinhard.lakes-harlan@uni-giessen.de

B. Hedwig (ed.), *Insect Hearing and Acoustic Communication*,
Animal Signals and Communication 1, DOI: 10.1007/978-3-642-40462-7_2,

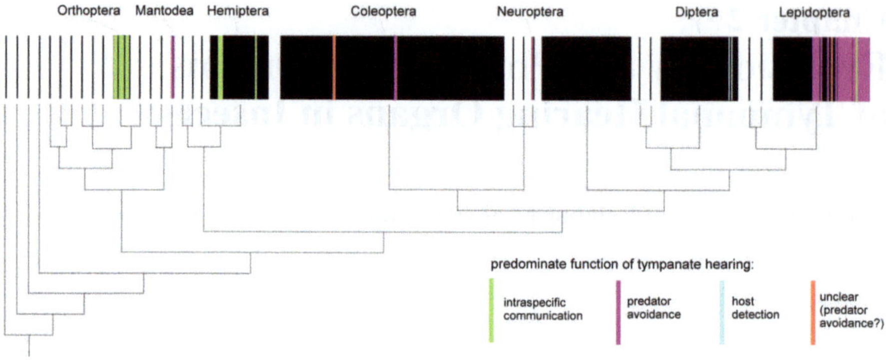

Fig. 2.1 Phylogeny of insect taxa as adapted from Wheeler et al. (2001) and Grimaldi and Engel (2005) indicating the presence of tympanal auditory organs. The top row depicts recent tympanate taxa with their relative number of species; the main or basic function of hearing is indicated by colour: for intraspecific communication, predator detection or host detection. In these taxa, except for Neuroptera hearing organs evolved more than once. Secondary reduction of hearing also occurred but has not been depicted

three major behavioural contexts: (1) intraspecific communication e.g. mate attraction, courtship and rivalry behaviour, (2) detection and avoidance of predators and (3) host localisation by parasitoid insects.

Communication systems based on species-specific acoustic signalling evolved repeatedly among animals (Webster et al. 1992; Yager 1999; Christensen-Dalsgaard and Carr 2008). Sound signals are effective means for signalling and intraspecific communication as they are independent of the photophase, provide directional cues in cluttered environments and may convey the sex or behavioural state of the sender. The complex temporal pattern and frequency content of sound signals represents an important pregametic isolation mechanism (Pollack 2000; Greenfield 2002). Intraspecific acoustic communication is well described in Orthoptera and Hemiptera (Fig. 2.1) (see Chap. 6 by Greenfield on Lepidoptera). In many nocturnally flying insects (Mantodea, Neuroptera, Lepidoptera; Fig. 2.1) auditory systems are used to detect and avoid predation by bats (reviews: Hoy 1992; Conner and Corcoran 2012; Yager 2012; see Chap. 5 by Conner). By contrast, only a few species of two taxa of Diptera possess ears for the highly specialised function of host detection (Fig. 2.1; see Chap. 4 by Hedwig and Robert).

For different insect taxa, typical tympanal ears are located on specific tagmata of the body (Fullard and Yack 1993; Hoy and Robert 1996; Yager 1999; Stumpner and von Helversen 2001; Yack 2004). Auditory organs with tympana evolved at least 18 times independently in diverse taxa of seven insect orders as: in butterflies and moths (Lepidoptera), locusts, crickets and bush crickets or katyids (Orthoptera), flies (Diptera), cicadas and water striders (Hemiptera), beetles (Coleoptera), mantids (Mantodea) and lacewings (Neuroptera) (Fig. 2.1). The auditory systems can usually be assigned to a main behavioural function, although hearing in a species might be involved in more than one of these behaviours, e.g.

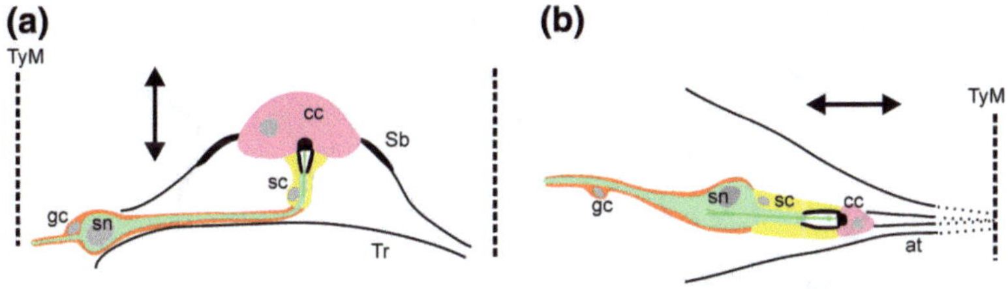

Fig. 2.2 Schematic scolopidal unit from a tettigoniid (after Schumacher 1979) and a diptera (after Lakes-Harlan et al. 2007). Each unit consists of a sensory neuron (sn, *mint*), glial cell (gc, *orange*), scolopidial cell (sc, *yellow*) and a cap cell (cc, *pink*). The proposed direction of sensory activation (*arrow*) in the tettigoniid is perpendicular to the tracheal wall and the unit is attached to a supporting band (sb) holding the cap cell. In the diptera, the tympanum is attached longitudinally to the sensory organ (at, attachment tissue). *TyM* tympanal membrane (not to scale)

in both mate recognition and predator detection (Moiseff et al. 1978; Wyttenbach et al. 1996; Rodriguez and Greenfield 2004).

Despite the different positions and forms of tympanal ears, shared functional elements in ears have become evident across taxa: tympanal ears are characterised by tympanal membranes or ear drums of thinned cuticle which vibrate in response to airborne sounds. The tympanum is typically backed by a tracheal space. There are notable exceptions of ears which are not backed by an acoustic trachea but by fluid, like the ear of the green lacewing (Miller 1970) and of some aquatic hemipterans (Arntz 1975). In some cases, an auditory trachea transfers the sound signal to the inside of the body. The third and crucial element is a sensory organ. This consists of scolopidial sensory units which are directly or indirectly activated by sound-induced tympanum oscillations and function as auditory receptor neurons (Hoy and Robert 1996; Yager 1999; Yack 2004). Scolopidial sensilla (Fig. 2.2a, b) are widespread among insects in proprioceptive and vibration-sensitive receptor organs (Moulins 1976; Field and Matheson 1998). Each scolopidial sensillum consists of four distinct cell types, only one of which is a primary sensory neuron. The dendrite of the sensory neuron extends distally into a cilium; the tip of the cilium is covered by a specific cap cell (this cap cell characterises the mononematic type of scolopidia, irrespective whether tympana are developed or not). Dendritic cilia of homogenous diameter are called type 1 scolopidia (Field and Matheson 1998; Yack 2004). Additional cell types of the scolopidium are the glia (sheeth) cell and scolopale cell. The latter forms a rod-like structure, the scolopale, which surrounds the dendrite. Within the scolopale, the dendrite is surrounded by fluid which may be secreted and homeostatically regulated by the scolopale cell. In tympanal ears, the dendrite of a single sensory neuron resides in the scolopale (monodynal type) and the ears have mononematic type 1 scolopidia (Yack 2004). In auditory systems, the sensory neurons are ultimately activated by sound stimuli which mechanically stretch the dendrites of sensory neurons (Kernan 2007).

The different evolutionary origins and functions correspond to a great diversity of ear morphologies (see Fig. 2.3 for anatomy of selected insect ears). The differences are also reflected in the varying number of scolopidial sensilla and their organisation.

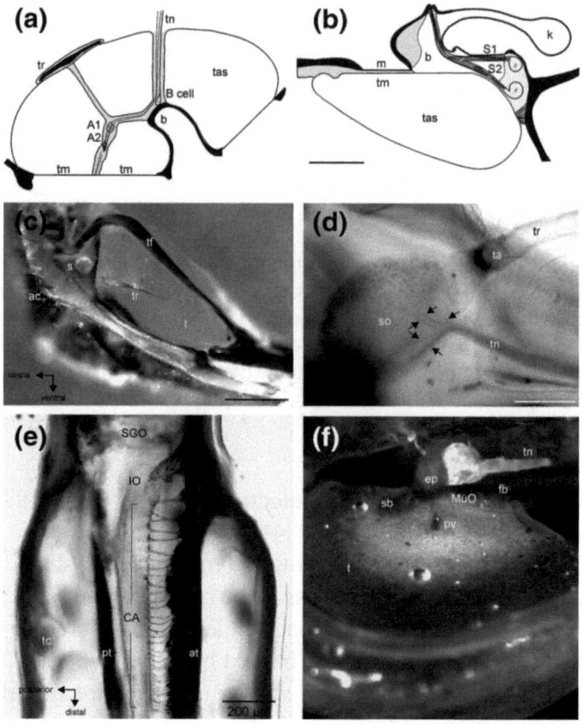

Fig. 2.3 Morphological diversity of insect tympanal ears. **a** Schematic transvers section of the tympanal organ of noctuoid moths containing only two transversscolopidial sensory neurons (A1, A2) attached to the tympanal membrane (tm). Axons of the sensory neurons form the tympanal nerve (tn). Another sensory cell, the B cell, is not directly connected to the tympanum. The tympanum is backed by tracheal air sacs (tas) and a trachea (tr). **b** Schematic section through the mesothoracic tympanal organ of a waterboat man, *Corixa punctata* (Heteroptera). Two sensory neurons (S1, S2) are connected to the tympanal membrane (tm) via a knob-shaped process (k) which upon tympanum movement stretches the dendrites. Sound input is delivered to sensory cells from the tracheal sac (t). Abbreviations: b, base of knob; m, membrane; S1/S2, sensory neuron 1/2; t, tracheal sac; tm, tracheal membrane. **c** External view of the tympanal organ of the cicada, *Okanagana rimosa*. The tympanum (t) is surrounded by a tympanal frame (tf) and contains a thickened tympanal ridge (tr). Close to a spiracle (s), the auditory capsule (ac) harbours the sensory organ, which is connected to the tympanal ridge by the tympanal apodeme (ta). **d** Internal view of the sensory organ (so) of *O. rimosa*. Distinct minor nerve branches of the tympanal nerve (tn) run within the mass of the sensory organ (arrows). Sensory cells are coupled via the tympanal apodeme (ta) to the tympanal ridge (tr), which transfers the resonant vibrations of the tympanum. **e** The complex tibial organ in the foreleg of a tettigoniid *Tettigonia viridissima* containing three scolopidial organs, the subgenual organ (SGO), intermediate organ (IO) and *crista acoustica* (CA) located behind the tympana (at, anterior tympanum, pt, posterior tympanum) in tympanal chambers (tc). Note the parallel organisation of dendrites of auditory CA neurons. **f** Internal view of the tympanum and auditory organ (Müllers organ, MüO) of the desert locust *Schistocerca gregaria*. Auditory receptors have different attachment sites directly on the tympanal membrane: the folded body (fb), ep (elevated process), pyriform vesicle (pv), and styliform body (sb). **a** redrawn from Roeder (1967), **b** redrawn from Prager (1976), **c–f** Strauß and Lakes-Harlan unpublished. Scales: **b** 100 μm, **c** 1 mm, **d** 400 μm and **e** 200 μm

Insect ears may contain only one or two scolopidial sensilla like in moth (Fig. 2.3a) or water striders (Fig. 2.3b) or may contain over 2000 as in cicadas (Doolan and Young 1981) (Fig. 2.3c, d). The orientation of dendrites and the attachment of sensory neurons to the tympana or trachea vary greatly in different ears. Frequency analysis may be based on attaching the sensory cells specifically to sound-transmitting structures like in the ears of tettigoniids where receptor cells are linearly arranged (Fig. 2.3e) or in locusts where distinct neuron groups attach to specific areas of the tympana (Fig. 2.3f). Comparative analysis of auditory structure and function has recently expanded the understanding of evolutionary origins of insect ears both empirically and conceptually. In the following, the evolution of tympanal sound pressure receivers for different tasks is reviewed in relation to phylogenetic analysis.

2.2 Comparative and Phylogenetic Studies of Tympanal Organs

Several sources of information are relevant to understand the evolution of tympanal organs. Descriptive data on the neuroanatomy and functional morphology of tympanal organs analyse sensory organs in different species. Comparative and phylogeny-based studies are based on relationships of organisms. These latter studies aim to reconstruct the evolutionary sequences of e.g. morphological and functional changes. The concepts of cladistics are used to identify the ancestral (plesiomorphic) lineages and anatomical features of sense organs, or the derived (autapomorphic) situations (so-called "tree-thinking"). Phylogenetic analysis indicates in which lineage tympanal organs evolved or became secondarily reduced and may point out suitable recent taxa for understanding evolutionary transitions. Reconstructing phylogenetic relationships by molecular techniques has provided numerous phylogenies for insect taxa, which however, have so far not resolved evolutionary processes for all taxa with tympanal ears. Combined approaches including neuroanatomy, behaviour, and physiology of auditory systems should provide important characters for a phylogenetic analysis, which may also allow identifying the selection pressures acting upon during evolution.

2.2.1 Fossil Record of Insect Ears

The fossil record for insect ears is scarce and only known from Ensifera (Orthoptera) (Rust et al. 1999; Plotnick and Smith 2012) and date back to the Eocene (~56–34 million years ago). Some well preserved specimen of gryllids and tettigoniids from this era (dating at ~48 my) show tympana reminiscent of the tympanal ears of recent species (Plotnick and Smith 2012). By that time, intraspecific acoustic communication was probably long established. Fossils from the wing stridulation apparatus allow even to reconstruct the ancient sounds. Analysis of stridulation teeth distance in the Jurassic (165 my ago) haglid *Archaboilus musicus* suggests a carrier frequency of 6.4 kHz in their pure-tone song (Gu et al. 2012). Songs from the tettigoniid *Pseudotettigonia amoena,* in the Eocene (55 my ago), were likely broad-banded, with a carrier frequency around 7 kHz (Rust et al. 1999). Importantly, an ancestral bat species dated to the early Eocene at ~52 my shows no indications of ultrasonic sound production (Simmons et al. 2008). Thus, hearing in Ensifera likely evolved prior to bat echolocation, and its ancestral function was presumably linked to intraspecific communication (Stumpner and von Helversen 2001).

2.2.2 Comparative Neuroanatomical Approaches

Comparative anatomical studies compare sensory organs of related species (interspecific comparison), or serially homologous organs, i.e. sensory organs occurring in similar positions in different body segments of the same species (intraspecific comparison).

In interspecific comparisons the homologous sensory structures in hearing (tympanate) species may be compared to closely related recent atympanate species (Orthoptera: Meier and Reichert 1990; Lepidoptera: Yack and Fullard 1990; Yack et al. 1999, Mantodea: Yager 2005; Diptera: Edgecomb et al. 1995; Lakes-Harlan et al. 2007) assumed to be representative of the ancestrally deaf situation (Fullard and Yack 1993) though they might also result from secondary reduction (see below). Homology is usually evaluated based on the morphological analysis of adult ears or their development (Yager 1999). To identify sensory organs in non-hearing species which are homologous to ears in related species, the three morphological criteria of homology are to be met. These criteria are those of location (homotopy), the special quality of organisation (homomorphy), and of continuity (either in several species, thus "linking" morphological forms, or during ontogenetic development of one species). Homology analysis of ears is usually based on the location and quality of organisation.

Comparative studies provided considerable insights into the origin of insect ears and importantly have identified homologous sensory organs in tympanate and atympanate species. Auditory sense organs are derived from pre-existing sensory organs which served functions other than hearing, e.g. proprioception (see below). The lineage-specific anatomical structures indicate which changes accompanied the transformation of ancestral non-hearing mechanosensory organs into ears. For example, the embryonic development of the locust (*Schistocerca gregaria*) ear indicated that the auditory sense organ evolved among scolopidial organs present in all abdominal segments (Meier and Reichert 1990; Schäffer and Lakes-Harlan 2001) and derived from a rather unspecialised, proprioceptive organ. The early development of the auditory sense organ and the serially homologous pleural chordotonal organs is highly similar. However, part of the neurons in the differentiating auditory sense organ migrate anteriorly during embryogenesis and generate the anterior group of receptors. Thereby they establish the morphology of the adult hearing organ which contains more neurons than the unspecialised scolopidial organs. Among tettigoniids, tympanal organs occur in the foreleg while homologous sensory organs develop in the mid- and hindleg without tympana (Rössler 1992). Adaptations for hearing are the tympanal membranes in the foreleg (Bailey 1993) and an acoustic spiracle and trachea as main sound entrance to the auditory system (Lewis 1974; Nocke 1975).

2.2.3 *Phylogeny-Based Approaches*

Given a reliable phylogenetic tree, neurosensory data can be projected onto it in order to deduce the most likely sequence of evolution. Such neurosensory data concern the anatomy of the scolopidial organs, their innervation, their serial organisation, the sensory attachment, or ultrastructural features. In combination with the species' phylogenetic relationships, it is possible to infer in which lineage(s) and therefore in which succession, the corresponding changes in sensory structures evolved. A phylogenetic framework can become particularly important for the identification of two distinct evolutionary processes: secondary reduction of tympana, and their parallel evolution. If tympana have been secondarily reduced, insects might be mistaken for primarily atympanate. Parallel evolution in independent lineages has resulted in ears with high structural similarity based on the same ancestral organ. One example of parallel evolution concerns the ears of flies. Among Diptera, tympanal hearing evolved only in two groups, the parasitoid Emblemasomatini and Ormiini, whose ears are located in the prothorax. As both taxa are not closely related phylogenetically, a parallel evolution of hearing can be deduced (Fig. 2.4) (Lakes-Harlan et al. 1999). Diptera possess a prothoracic chordotonal sense organ which twice has been modified and incorporated into an ear together with the development of a prosternal tympanal membrane

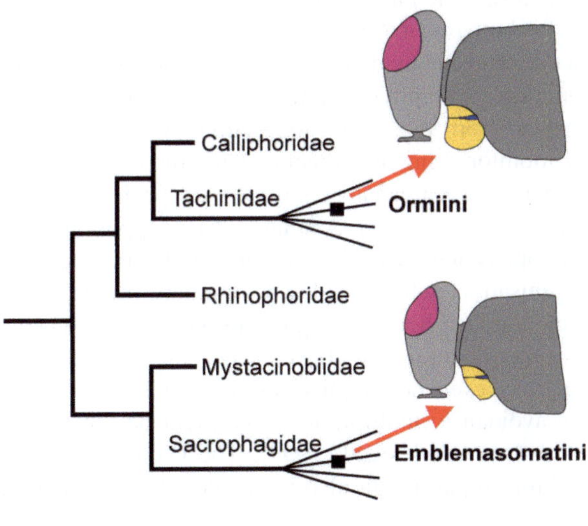

Fig. 2.4 Prothoracic tympanal organs in two lineages of calyptrate Diptera Emblemasomatini and Ormiini. This phylogenetic mapping supports independent evolution of the hearing organs. Adapted from Lakes-Harlan et al. (1999) with a phylogeny modified after Kutty et al. (2010)

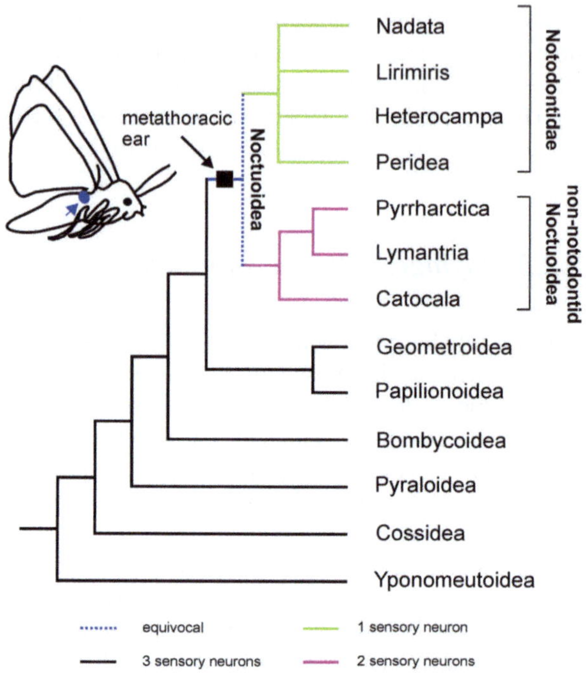

Fig. 2.5 Number of sensory neurons in the metathoracic ear of lepidopteran species. Tympanal ears evolved in the lineage of Noctuidea (*black square*). The number of scolopidial sensillae is indicated by colour. The ancestral number is three neurons, which was reduced to two neurons in most Noctuidea or one neuron in Notodontidae. Adapted from Yack et al. (1999), animal redrawn from Yack and Fullard (2000)

(Lakes-Harlan and Heller 1992; Edgecomb et al. 1995). The prosternal chordotonal organ is a vibration sensor (Stölting et al. 2007; see below). Consistent with a parallel evolution of hearing organs is the fact that the host species of these parasitoid flies belong to different taxa: Ormiini infest species of Orthoptera while Emblemasomatini infest cicadas (see Chap. 4 by Hedwig and Robert). However, the physical conditions and constraints making the scolopidial organ and structures a suitable base are unclear.

In the case of Macrolepidotera, a reduction of the number of sensory neurons in ears became evident by blending comparative neuroanatomy of diverse lepidopteran taxa with phylogenetics. The metathoracic tympanal ears in some Lepidoptera derived from the wing-hinge chordotonal organ, an (atympanate) receptor organ to monitor wing movements with three scolopidial neurons (Yack and Fullard 1990). The tympanal ears of recent species contain either two or just one single sensory neuron (Figs. 2.3a and 2.5). Mapping the neuroanatomical details onto the phylogeny identifies the situation with three neurons as the ancestral atympanate organisation, while the lower numbers reveal an evolutionary reduction: ears of Notodontidae house a single sensory unit (Surlykke 1984; Yack et al. 1999), while all other noctuoid ears have two (Yack et al. 1999). The reduction in neuron number may even be adaptive, as noctuoids do not use frequency discrimination for bat avoidance; multiple neurons might be functionally unnecessary, and would be redundant (Yack et al. 1999). The evolutionary transition from proprioceptor to auditory organ might also implicate a loss of the original function but it is unclear how this loss is compensated. Reduction of tympana or at least hearing sensitivity has been documented in Lepidoptera (Cardone and Fullard 1988; Fullard et al. 2007) Mantidae (Yager 1990), Orthoptera (Otte 1990; Pollack and Martins 2007; Lehmann et al. 2010) and is often associated with a regression or loss of wings, indicating that flying insects are under particular selection pressure from echolocating bats (Yager 2012). If the ability of flight is lost, the selection on hearing may decrease so that ears are undergoing regression and might also be lost.

Phylogenetic mapping and any inference drawn crucially depend on the quality of the phylogenetic analysis available and its resolution. When relationships between taxa and identification of outgroups were not (yet) reliably established, improvement of phylogenies resulted in new concepts of ear evolution. As mentioned above, ultrasonic hearing evolved repeatedly among Lepidoptera, with ears in different positions on the body, including metathoracic and abdominal ears (Yack and Fullard 2000; Yack 2004). Among three groups of Lepidoptera (Pyraloidea, Geometridae and Drepanidae), abdominal ears appeared to have evolved in parallel (independently) from a lateral scolopidial organ in the first abdominal segment (Hasenfuss 1997). As the relationships among Lepidoptera was recently revised based on molecular sequence data, the parallel evolution of hearing organs was questioned and instead a singular evolution of ears discussed (Regier et al. 2009). However, the singular evolution of ears in the lineage of all tympanate groups can be ruled out based on morphological analysis of homology as ears are located in different segments (Kristensen 2012). A similar incongruence occurs for Ensifera as morphological and molecular phylogenetic analysis (Legendre et al. 2010) lead to different scenarios regarding the evolution or reduction of tympanal ears. In general, the homology of ears and especially their sensory organs has to be established independently of the phylogenetic relationships. This neuroanatomical approach is crucial for homology analysis, as sensory organs which are not homologous cannot share a common evolutionary origin.

The examples discussed so far have all used given phylogenies to discuss the evolution of hearing organs. However, auditory features can become an integral part of a phylogenetic study if included in the cladistic analysis. The most prominent example so far is hearing in Mantidae (Yager and Svenson 2008). The mantid ear is located at the ventral metathorax and forms an "auditory cyclops". Based on functional anatomy and neurophysiology, five distinct ear types can be distinguished in the metathorax of mantids. While the ear shows some variability in morphology, the sensory physiology and behavioural response to ultrasound are remarkably similar. Using different structural features a morphological data matrix was created and an "auditory phylogeny" generated (Yager and Svenson 2008). Outgroup cockroach species as well as mantid instars, which lack fully developed auditory systems, clustered at the base. The topography of the phylogenetic tree supported a single origin of the metathoracic ear, while another hearing organ in the mesothorax with auditory sensitivity to frequencies lower than ultrasound may have evolved repeatedly. Comparison with mantid phylogeny based on molecular sequences supports that the metathoracic ear evolved once. In addition, a distinct type of the metathoracic ears which is rather unelaborate as it lacks clearly defined acoustic chambers is not the ancestral ear type, but developed repeatedly (Yager and Svenson 2008) and may represent a regression associated with wing reduction. Thus, neurosensory characters can provide a data set for reconstructing ear evolution using cladistic approaches. Analysis of, e.g., hearing organs and their organisation could therefore be included for cladistic analysis to distinguish homology from convergence (e.g. Desutter-Grandcolas et al. 2005), especially if ear homology is contentious.

Neurobiology and phylogeny are both important to reconstruct the evolution of tympanal ears. Phylogeny alone cannot sufficiently explain how the tympanal ears of insects formed; for this the function of the ancestral receptor organ and its evolutionary modification has to be considered. Phylogenies are decisive to distinguish between monophyly versus polyphyly of hearing, but only data on the sensory organs can reveal the direction of anatomical change by identifying the ancestral structure of sensory organs.

2.3 The Evolutionary Origin of Auditory Sense Organs

Scolopidial sensilla occur in tympanal organs, but moreover in numerous organs of proprioceptive or vibroceptive function in insects (Field and Matheson 1998). Scolopidial organs in similar locations and even with similar organisation to auditory sensilla are often present in related tympanate and atympanate species. As the sensory structures and the mechanism of activation are conserved among proprioceptive, vibroceptive and auditory organs (Kernan 2007), insect ears apparently evolved from pre-existing mechanoreceptor organs (Meier and Reichert 1990; Boyan 1993; Fullard and Yack 1993; Yager 1999) which were termed "precursor organs" (Yager 1999). This concept of a precursor organ as basis of the evolution of ears was conceived already in the nineteenth century (Graber 1881). However, the idea of a mechanoceptive origin of auditory organs usually addressed only specific taxa of insects (but see Radl 1905), and it took some time to identify precursor organs and their ancestral function for several insect ears. According to the precursor concept, the scolopidial sensilla were already present prior to an auditory function; only the medium for effective stimulation changed by acquiring sensitivity for airborne sound. For any scolopidial organ at least two criteria have to be met to identify it as an ear precursor: (1) It has to be homologous to the sensory organ of tympanate taxa as derived from morphological analysis. (2) A phylogenetic analysis has to establish that the tympanate and atympanate taxa share a common atympanate ancestry which the recent atympanate species might still represent. An alternative scenario would be that ears evolved by generating additional sensory organs ("de novo") which directly evolved as the auditory sense organ. Though this is in principle a valid possibility, it has so far not been supported for any insect ear. Apart from location, precursor organs also match in their innervation, their axonal projection to mechanosensory areas of the CNS, and their synaptic connections to interneurons in atympanate and tympanate species (Yack and Fullard 1990; Boyan 1993;

Table 2.1 Chordotonal precursor organs of insect ears

Taxon	Position of auditory organ	Precursor	Reference
Diptera			
Sarcophagidae	Prothorax	Prosternal CO	Lakes-Harlan et al. (1999)
Tachinidae	Prothorax	Prosternal CO	Lakes-Harlan et al. (1999)
Coleoptera			
Cicindelidae	1st abdominal segment	Pleural CO	Yager and Spangler (1995)
Acrididae	First abdominal segment	Pleural CO	Meier and Reichert (1990)
Ensifera			
Grylloidea	Foreleg tibia	Intermediate organ	Strauß and Lakes-Harlan (2009)
Tettigonioidea	Foreleg tibia	Crista acustica homolog	Strauß and Lakes-Harlan (2009)
Mantodea	Ventral metathorax	N7 chordotonal organ	Yager (2005)
	Ventral mesothorax	N7 CO of mesothorax	
Lepidoptera			
Sphingidae	Pilifer	Single SO	Göpfert and Wasserthal (1999)
Noctuoidea	Metathorax	Wing-hinge CO	Yack and Fullard (1990)
Pyraloidea	Ventral 1st abd segment	Lateral scolopal organ	Hasenfuss (1997)
Geometridae	Anterior 1st abd segment	Lateral scolopal organ	Hasenfuss (1997)
Drepanidae	1st abdominal segment	Lateral scolopal organ	Hasenfuss (1997)
Uraniidae	2nd abdominal segment	Lateral CO	Hasenfuss (2000)

CO chordotonal organ; *SO* scolopidial organ

Yager 1999). The precursor organs of ears have been identified for several taxa, but not yet in Hemiptera, Coleoptera (scarab beetles) and some Lepidoptera (in Hedylidae and Nymphalidae) (Table 2.1).

Evolution of tympanal ears was apparently rather complex in Ensifera. Gryllids, tettigoniids and few related taxa have tympanal ears within the tibia of the forelegs with usually two tympana in the anterior and posterior tibia, respectively (Fig. 2.3e). In the atympanate Rhaphidophoridae (cave crickets), the subgenual organ and the intermediate organ are present, but no sensory neurons homologous to the auditory receptors (Jeram et al. 1995). This organisation may well be the ancestral atympanate situation of Ensifera. In the tettigoniids and gryllids, the hearing organs are similar in neuroanatomical structures, but important differences exist. In tettigoniids, wetas (Anostostomatidae) and haglids, three sensory organs form the complex tibial organ: the subgenual organ, the intermediate organ, and the sensory neurons in the *crista acustica* responding to airborne sound. In a majority of the non-hearing Ensifera including Stenopelmatidae and Gryllacrididae, sensory structures are present in the complex tibial organ which is clearly homologous to the

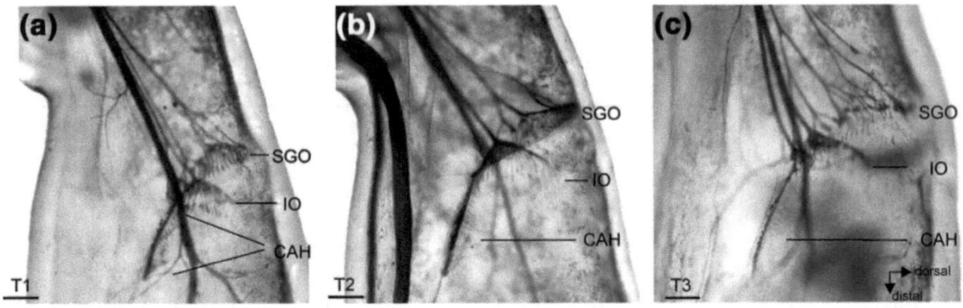

Fig. 2.6 The complex tibial organ of *Stenopelmatus spec.* in the thoracic leg pairs (T1; foreleg, T2; midleg, T3; hindleg). In this species, no tympana are developed in the foreleg but the *crista acustica* homolog corresponds to the auditory sensilla of related tettigoniid groups. No differences in the neuronal organisation are evident between leg pairs. *CAH* crista acustica homolog; *IO* intermediate organ; *SGO* subgenual organ. Scale: 1 mm

auditory receptors in tettigonioids (Fig. 2.6). As they resemble the *crista acustica* of hearing taxa, they have been termed *crista acustica homolog* (CAH) (Strauß and Lakes-Harlan 2008a, b, 2010). The sensory organ of the *crista acustica homolog* is highly similar between all leg pairs and shows no sensory specialisations relating to hearing in the foreleg; therefore it is assumed to represent the ancestral and atympanate organisation. For several of these atympanate Ensiferan taxa, which do not use acoustic signals for intraspecific signalling, communication with vibratory signals plays an important role in intraspecific behaviour (e.g. Field and Bailey 1997; Weissman 2001; Gwynne 2004). Many insects perceive vibration signals with highly sensitive subgenual organs which in Ensifera are also part of the complex tibial organ besides the atympanate *crista acustica homolog*, which might have developed with a role in vibration detection (Strauß and Lakes-Harlan 2009; 2010). The *crista acustica homolog* in atympanate taxa might thus be the precursor organ for the *crista acustica* in Tettigonioidea (Fig. 2.7).While the phylogeny of Ensifera is still debated, the neuroanatomy of sense organs suggests a consistent scenario how ears mayhave originated amongst Ensifera (Fig. 2.7). Improved taxon sampling and phylogenetic analysis should help to delineate the sensory adaptations.

The functional changes in a mechanosensory organ are also interesting in respect to the upstream sensory pathways in the central nervous system: in tettigoniids and grasshoppers, support for a vibratory origin of the auditory system is given by many interneurons with bimodal vibratory and auditory response properties (Kalmring and Kühne 1980; Kalmring et al. 1997). In this case, audition seems to be an evolutionary addition onto the vibration-sensitive system.

In the grylloid group with e.g. crickets and mole crickets, ears also occur in the foreleg but the neuroanatomy of the complex receptor organ differs from that in tettigoniids. It consists of only two main organs, the subgenual organ and the tympanal organ, the latter being the auditory receptor cells proper. Most likely, this tympanal organ does not correspond to the *crista acustica*. Presumably, the cricket tympanal organ derived from the intermediate organ or distal organ found in many

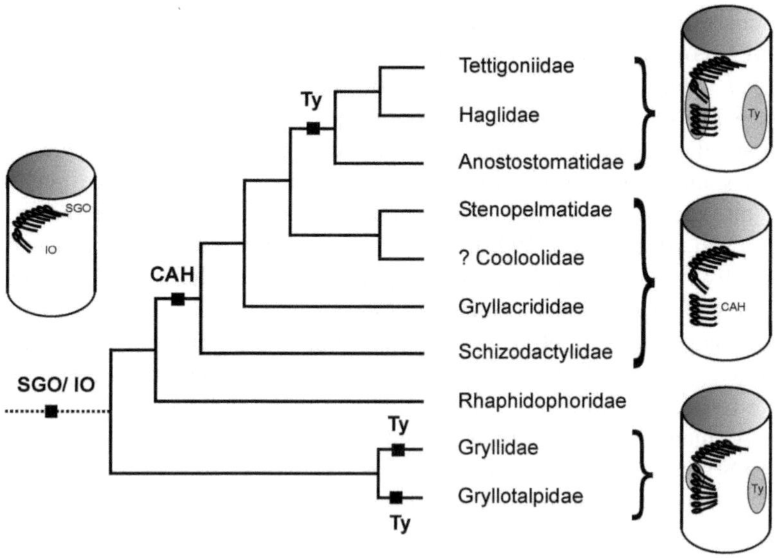

Fig. 2.7 Evolution of tympanal ears in Ensifera. The complex tibial organ consisted ancestrally of the subgenual organ (SGO) and the intermediate organ (IO). In the tettigoniid lineage a third part evolved, the *Crista acustica homolog* (CAH). Tympana (tymp) occur only later to form an auditory organ. In the grylloid lineage, tympanal hearing organs evolved presumably repeatedly from a dipartite sense organ, the plesiomorphic intermediate organ. In the Cooloolidae, the neuroanatomy of the ears is not yet known. Phylogeny based on Desutter-Grandcolas (2003); neurosensory data after Strauß and Lakes-Harlan (2009)

taxa. Remarkably, a much greater anatomical diversity exists for ears in crickets than in tettigoniids, including an ear type in *Cycloptiloides canariensis* (Gryllidae: Mogoplistinae) with a single anterior tympanum and very few scolopidia (Michel 1979). Phylogenetic analysis even suggested the independent origin of tympana in crickets and mole crickets (Desutter-Grandcolas 2003). Presumably, the hearing organs in gryllid and tettigoniid lineages evolved independently, and from different sets of sensilla.

2.4 The Evolutionary Origin of Physiological Hearing

Identification of the precursor organ reveals from which structure the sensory organ derived. With respect to its functional evolution, additional information on accompanying physiological changes and the ecological context of auditory behaviour are necessary.

During the evolutionary transformation of scolopidial organs to functional ears structural changes occur which alter the range and quality of stimuli to which the chordotonal organ responds. The sensory precursor systems may have been sensitive to high amplitude sound without any specific sound-propagating structures. For example, in cockroaches, the subgenual organ is sensitive to low-frequency sound (Shaw 1994). Similarily, the hind wing chordotonal organ of locusts reacts to sound, beside its proprioceptive function (Pearson et al. 1989). It appears that precursor organs were sensitive to high amplitude sounds only, and that sound-propagating structures evolved in consecutive steps transforming cuticle into the tympana and in some groups a respiratory trachea into an acoustic trachea (Fullard and Yack 1993). Accordingly, atympanate sensory organs which are (serially) homologous to auditory organs have been shown to respond to high amplitudes of low-frequency sounds in Lepidoptera (Yack and Fullard 1990) and Tettigoniidae (Kalmring et al. 1994; Jeram et al. 1995). This suggests that during evolution a gain of auditory sensitivity occurred in a continuum of sensory activity.

With respect to auditory sensilla numbers it is difficult to identify evolutionary trends. As mentioned above, in Noctuoidea, a decrease of sensilla occurred (Yack et al. 1999), however, in other Lepidoptera, the number of sensilla did not change during the evolution of tympanal hearing from the precursor organ (Hasenfuss 2000). In general, it might be assumed that sensory neurons in tympanal auditory organs may have been added to enhance sensitivity and frequency range. In insect taxa with intraspecific acoustic communication, the number of auditory afferents can vary almost over two orders of magnitude. Cicada which process very distinct and complex frequency modulated song patterns (Fonseca et al. 2000) have ears with over 2000 sensory neurons (Doolan and Young 1981). Also the pneumorid Caelifera *Bullacris membracioides* (a Bladder grasshopper) house 2000 auditory afferents in their main—atympanate—hearing organ (van Staaden and Römer 1998). For cicadas and pneumorid grasshoppers the functional role of the large number of receptor neurons is not resolved and the ancestral situation has not been investigated. Ensifera with intraspecific acoustic communication have 15–70 auditory sensilla. In several tettigoniids, the number of sensory neurons is not markedly higher than that in the atympanate precursor (22–35 sensilla; Strauß and Lakes-Harlan 2008a, b). However, the number of auditory sensilla in tympanate species is always significantly higher in the foreleg. In the genus *Poecilimon*, all species have well developed ears, but in few species less than 25 auditory sensilla occur (Lehmann et al. 2007; Strauß et al. 2012). This suggests a secondary reduction and may correlate to an adaptive decrease in auditory sensitivity due to shortened communication distances or even parthenogenetic reproduction (Stumpner and Heller 1992; Lehmann et al. 2007).

In non-hearing flies, the atympanate scolopidial prosternal organ, which is homologous to the ear, responds to vibrations of the legs (Stölting et al. 2007). In tachinid ears, functional specialisations for auditory perception can be identified as a widening of trachea backing the ear, the formation of tympana, an increase in neuron number and the elaboration of sensory cell attachment to the tympanal membrane. Tympanate Tachinidae have several hundred sensory neurons (Edgecomb et al. 1995; Lakes-Harlan et al. 2007), whereas non-hearing sarcophagids and the tympanate sacorphagid *E. auditrix* posses about 30–35 scolopidial units (Lakes-Harlan et al. 1999; Stölting et al. 2007). In hawkmoths, the structure of ears indicates that hearing evolved independently in *Choerocampina* and *Acherontiina*. In both groups, ears locate on the pilifer, but the *Choerocampina* have developed tympana proper while the *Acherontiina* use overlapping scales to pick up mainly ultrasonic sound (Göpfert et al. 2002).

A central question is how the physiological properties of the precursor organ changed for the functional transition to an auditory organ proper, but remarkably few insights have been established so far. With the gain of auditory sensitivity, the chordotonal organ might alter or lose the ancestral function of vibroception or proprioception. Different solutions are possible:

1. The hearing organ might still be bifunctional, responding to auditory as well as vibration/stretch stimuli. For example, the dipteran ear might represent a bifunctional sensory organ as it responds to sound and to substrate vibrations transmitted over the fly's body as in atympanate species (Lakes-Harlan et al. 1999). The ear of the grass moth *Pleuroptya ruralis* is located ventrally in the first abdominal segment. Auditory neurons are not completely mechanically isolated as they are activated by muscles located in the ventral diaphragm which is indirectly coupled to the receptor organ (Hasenfuss 2000). In this case, the auditory neurons might also be functional proprioceptors.
2. The original function of the precursor is compensated for by other sense organs. In Ensifera, presumably the sense organs in the atympanate legs took over the task of vibration reception.
3. The ancestral sensory function might no longer be necessary. The praying mantis ear presumably evolved with a loss of propioceptive function (Yager and Svenson 2008), as it might be the case in the Noctuid moth.

2.5 The Evolutionary Origin of Auditory Behaviour

Driven by natural and sexual selection, insect ears evolved mainly for intraspecific communication and predator detection. Therefore, the behavioural and ecological context of auditory systems is important for understanding their evolution. The most prominent example is the "arms-race" between echolocating bats and nocturnal flying insects (reviews: Hoy 1992; Conner and Corcoran 2012; Yager 2012). Bat detecting ears should be highly sensitive to ultrasound (Stumpner and von Helversen 2001) and physiological data show that they are usually tuned to a broad ultrasonic spectrum (Yack and Hoy 2003). Loosing predation pressure results in partial regression of auditory systems (Fullard et al. 2007). In Cicadidae, intraspecific acoustic signalling derived from ancestral vibration signalling in Auchenorrhyncha (Claridge et al. 1999; Hoch et al. 2006) (Fig. 2.8). Hemiptera of small body size use ancestrally tymbal mechanisms to generate vibration signals and Cicadidae produce auditory sounds by identical means of tymbal mechanisms. Thus, the biomechanical way of signal production was retained, but the signal's mode of transmission changed, as did the sensory mode of reception. Tympanal hearing probably evolved in parallel to this change in transmission mode.

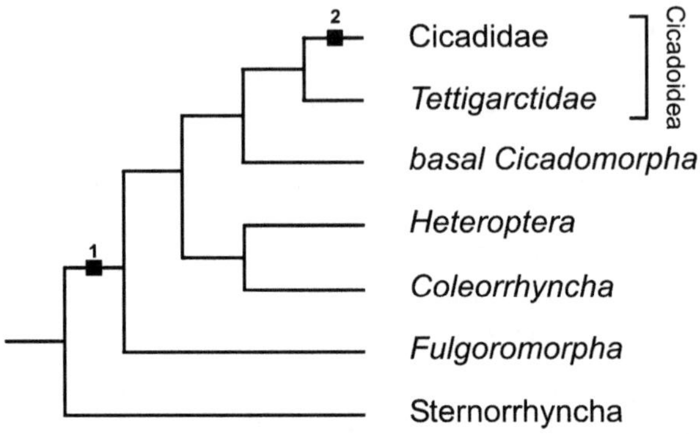

Fig. 2.8 Phylogenetic origin of cicada acoustic signalling revealed by mapping of signalling behaviours on the phylogeny of Hemiptera. Signal production by tymbals is present in all Hemiptera excluding the basal Sternorrhyncha, and thus presumably monophyletic (indicated as autapomorphy, a novel trait [1]. In ancestral taxa (names in italics) tymbals were used for the production of substrate vibration. Only recent Cicadidae use timbals for sound production and hear with tympanal ears (autapomorphy [2]). Their sister group, the Tettigarctidae, also produce vibration signals. Thus, acoustic communication evolved from tymbal mechanisms used for vibratory signalling. Adapted from Hoch et al. (2006) and from Claridge et al. (1999)

Acrididae commonly have ears but not all groups use acoustic signals for intraspecific communication, and those using auditory cues do so by different mechanisms (Riede 1987). A phylogenetic study indicated a possible order of the evolution of ears (Fig. 2.9): tympana are present in the Acridoidea, including grasshoppers and locusts, and the Pyrgomorphidae in the first abdominal segment while the latter lack stridulation behaviour. These two groups are sister taxa (Flook et al. 2000), making the assumption of an origin of tympana in their common lineage most parsimonous. Some more basal acridid taxa do use stridulation for sound production, though with different mechanisms than Acridoidea. It is thus difficult to match the origins of hearing and of stridulation, as for stridulation repeated gains as well as several losses may have occurred. Two explanations are available to accommodate these findings, conflicting over the ancestral function of hearing (Flook et al. 2000): stridulation and hearing evolved simultaneously within the lineage of Acridomorpha, and either one was lost in specific lineages. This implies that stridulation was initially also present in the lineage of Pyrgomorphidae, but was lost secondarily. As an alternative, the evolution of tympana and stridulation may have occurred independently, with audition presumably functioning in predator detection and evasion. Intraspecific signalling in Acrididae occurs by different sound production mechanisms, including mandibular, femoro-tegminal and femoro-abdominal sounds (Riede 1987). Even within a species, males and females might have different stridulatory structures, indicating their separate evolutionary origins, and hinting that stridulatory structures apparently evolve easily. Stridulation might have evolved in defence rather than intraspecific signalling. The multiple mechanisms of stridulation were also seen as indication that hearing

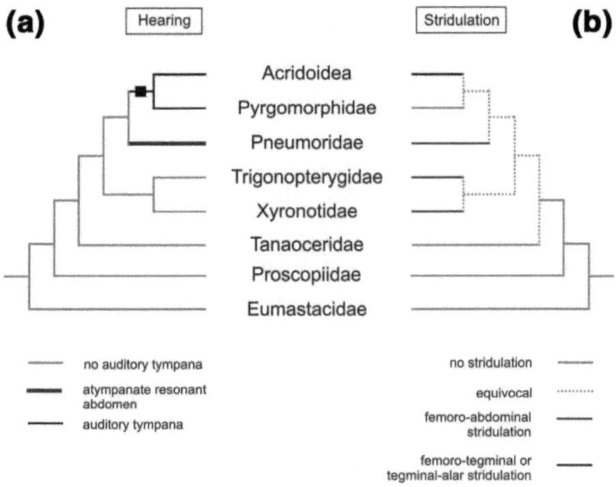

Fig. 2.9 Phylogeny of Acridomorpha with occurance of **a** tympanal organs and **b** stridulation mechanisms in the different taxa. Comparison of the distribution of hearing and sound production highlights their uneven distribution among Acridomorpha. As the most parsimonous explanation, tympanal hearing originated in the lineage of Acrioidea and Pyrgomorphidae (indicated by *square*). Origins of stridulation are not resolved, indicated by dotted line for equivocal taxa, and may involve repeated losses. The phylogenetic mapping highlights that stridulation occurred prior to hearing in either scenario in at least some taxa. Adapted from Flook et al. (2000)

evolved prior to intraspecific acoustic signalling (Riede 1987; Flook et al. 2000), probably to avoid terrestrial predators like reptiles (Stumpner and von Helversen 2001; Greenfield 2002) or small mammals (Flook et al. 2000). In this case, sound production came before sound perception by tympanal ears.

Amongst Noctuoidea and Pyraloidea moths, an evolutionary succession of acoustic functions seems to have taken place though in another sequence than in acridids. Some species produce intraspecific sound signals, but tympanal hearing is far more common. The evolution of intraspecific communication in this case seems to be facilitated by the hearing sense which evolved earlier in defence against bat predation. Male ultrasonic signalling may have been favoured in evolution due to the auditory system which was already sensitive to ultrasound (Greenfield 2002; see Chap. 6 by Greenfield).

2.6 Conclusion

The evolutionary approach to hearing in insects is a multidisciplinary one where quantitative behavioural analysis, behavioural ecology, and phylogenetics are just as important as functional anatomy and sensory physiology to understand the origin of insect ears. These studies address some central evolutionary questions, which all contribute to a proper understanding of hearing in insects: (1) from which sensory precursor organ did an ear evolve, (2) where in a given lineage did tympanal ears appear, (3) under which selection pressure(s) did the ear originate, (4) which physiological and behavioural adaptations occurred with a hearing sense, and (5) how did the hearing organ diversify further?

The depth of information is so far not the same for all tympanate taxa of insects. For cicada the ancestry of signalling behaviour has been resolved, but a precursor of the ear is not explicitly identified. However, for some groups, a very consistent scenario of ear evolution has been formed, e.g. for Mantids and Acridids, based on comparative neuroanatomy, physiology, phylogeny and studies of acoustic behaviour. Origins of auditory systems can be illuminated by phylogenetic approaches and "tree thinking", as they seek to clarify organismic relationships, identify ancestral and derived sensory organisations, and identify evolutionary transitions. These approaches will enforce multi-species studies, which will in turn provide principle insights into the commonalities and diversifications of auditory sensory evolution. Addressing different levels of evolutionary change, such as the origin of ears, the adaptive changes in existing ears, or changes in acoustic behaviour, in a surely cross-disciplinary approach will help to understand the evolution of insect ears.

References

Arntz B (1975) Das Hörvermögen von *Nepa cinerea* L. Zur Funktionsweise der thorakalen Scolopalorgane. J Comp Physiol 96:53–72

Bailey WJ (1993) The tettigoniid (Orthoptera: Tettigoniidae) ear: multiple functions and structural diversity. Int J Insect Morphol Embryol 22:185–205

Boyan GS (1993) Another look at insect audition: the tympanic receptors as an evolutionary specialisation of the chordotonal system. J Insect Physiol 39:187–200

Cardone B, Fullard JH (1988) Auditory characteristics and sexual dimorphism in the gypsy moth. Physiol Entomol 13:9–14

Christensen-Dalsgaard J, Carr CE (2008) Evolution of a sensory novelty: tympanic ears and the associated neural processing. Brain Res Bull 75:365–370

Claridge MF, Morgan JC, Moulds MS (1999) Substrate-transmitted acoustic signals of the primitive cicada, *Tettigarcta crinita* Distant (Hemiptera Cicadoidea, Tettigarctidae). J Nat Hist 33:1831–1834

Conner WE, Corcoran AJ (2012) Sound strategies: the 65-million-year-old battle between bats and insects. Ann Rev Entomol 57:21–39

Desutter-Grandcolas L (2003) Phylogeny and the evolution of acoustic communication in extant Ensifera (Insecta, Orthoptera). Zool Scripta 32:525–561

Desutter-Grandcolas L, Legendre F, Grandcolas P, Robillard T, Murienne J (2005) Convergence and parallelism: is a new life ahead of old concepts? Cladistics 21:51–61

Doolan JM, Young D (1981) The organization of the auditory organ of the bladder cicada, *Cystosoma saundersii*. Phil Trans R Soc Lond B 291:525–540

Edgecomb RS, Robert D, Read MP, Hoy RR (1995) The tympanal organ of a fly: phylogenetic analysis of its morphological origin. Cell Tissue Res 282:251–268

Field LH, Bailey WJ (1997) Sound production in primitive Orthoptera from Western Australia: sounds used in defence and social communication in *Ametrus* sp. and *Hadrogryllacris* sp. (Gryllacrididae: Orthoptera). J Nat Hist 31:1127–1141

Field LH, Matheson T (1998) Chordotonal organs of insects. Adv Insect Physiol 27:1–228

Flook PK, Klee S, Rowell CHF (2000) Molecular phylogenetic analysis of the Pneumoroidea (Orthoptera, Caelifera): molecular data resolve morphological character conflicts in the basal Acridomorpha. Mol Phyl Evol 15:345–354

Fonseca PJ, Münch D, Hennig RM (2000) How cicadas interpret acoustic signals. Nature 405:297–298

Fullard JH, Yack JE (1993) The evolutionary biology of insect hearing. Trends Ecol Evol 8:248–252

Fullard JH, Ratcliffe JM, ter Hofstede H (2007) Neural evolution in the bat-free habitat of Tahiti: partial regression in an anti-predator auditory system. Biol Lett 22:26–28

Göpfert MC, Wasserthal LT (1999) Auditory sensory cells in hawkmoths: identification, physiology and structure. J Exp Biol 202:1579–1587

Göpfert MC, Surlykke A, Wasserthal LT (2002) Tympanal and atympanal 'mouth-ears' in hawkmoths (Sphingidae). Proc Roy Soc Lond B 269:89–95

Graber V (1881) Über die stiftführenden und chordotonalen Sinnesorgane bei den Insekten. Zool Anz 4:450–453

Greenfield MD (2002) Signalers and receivers: mechanisms and evolution of arthropod communication. Oxford University Press, Oxford

Grimaldi D, Engel MS (2005) Evolution of the insects. Cambridge University Press, Cambridge

Gu J–J, Montealegre-Z F, Robert D, Engel MS, Qiao G-X, Ren D (2012) Wing stridulation in a Jurassic katydid (Insecta, Orthoptera) produced low-pitched musical calls to attract females. Proc Natl Acad Sci USA 109:3868–3873

Gwynne DT (2004) Reproductive behavior of ground weta (Orthoptera: Anostostomatidae): drumming behavior, nuptial feeding, post-copulatory guarding and maternal care. J Kansas Entomol Soc 77:414–428

Hasenfuss I (1997) Precursor structures and evolution of tympanal organs in Lepidoptera (Insecta, Pterygota). Zoomorphology 117:155–164

Hasenfuss I (2000) Evolutionary pathways of truncal tympanal organs in Lepidoptera (Insecta: Holometabola). Zool Anz 239:27–44

Hoch H, Deckert J, Wessel A (2006) Vibrational signalling in a Gondwanan relict insect (Hemiptera: Coleorrhyncha: Peloridiidae). Biol Lett 2:222–224

Hoy RR (1992) The evolution of hearing in insects as an adaptation to predation from bats. In: Webster D, Fay R, Popper A (eds) Comparative and evolutionary biology of hearing. Springer, New York, pp 115–130

Hoy RR, Robert D (1996) Tympanal hearing in insects. Ann Rev Entomol 41:433–445

Jeram S, Rössler W, Cokl A, Kalmring K (1995) Structure of atympanate tibial organs in legs of the cave-living Ensifera, *Troglophilus neglectus* (Gryllacridoidea, Raphidophoridae). J Morphol 223:109–118

Kalmring K, Kühne R (1980) The coding of airborne-sound and vibration signals in bimodal ventral-cord neurons of the grasshopper *Tettigonia cantans*. J Comp Physiol 139:267–275

Kalmring K, Rössler W, Unrast C (1994) Complex tibial organs in the forelegs, midlegs, and hindlegs of the bushcricket *Gampsocleis gratiosa* (Tettigoniidae): comparison of the physiology of the organs. J Exp Zool 270:155–161

Kalmring K, Sickmann T, Jatho M, Zhantiev R, Grossbach M (1997) The auditory-vibratory sensory system of *Polysarcus denticauda* (Phaneropterinae, Tettigoniidae): III. Physiology of the ventral cord neurons ascending to the head ganglia. J Exp Zool 279:9–28

Kernan MJ (2007) Mechanotransduction and auditory transduction in *Drosophila*. Pflugers Arch Eur J Physiol 454:703–720

Kristensen NP (2012) Molecular phylogenies, morphological homologies and the evolution of moth 'ears'. Syst Entomol 37:237–239

Kutty SN, Pape Th, Wiegmann BM, Meier R (2010) Molecular phylogeny of the Calyptratae (Diptera: Cyclorrhapha) with an emphasis on the superfamily Oestroidea and the position of Mystacinobiidae and McAlpine's fly. Syst Entomol 35:614–635

Lakes-Harlan R, Heller K-G (1992) Ultrasound-sensitive ears in a parasitoid fly. Naturwissenschaften 79:224–226

Lakes-Harlan R, Stölting H, Stumpner A (1999) Convergent evolution of insect hearing organs from a preadaptive structure. Proc Roy Soc Lond B 266:1161–1167
Lakes-Harlan R, Jacobs K, Allen G (2007) Comparison of auditory sense organs in parasitoid Tachinidae (Diptera) hosted by Tettigoniidae (Orthoptera) and homologous structures in an non-hearing Phoridae (Diptera). Zoomorphology 126:229–243
Lehmann GUC, Strauß J, Lakes-Harlan R (2007) Listening when there is no sexual signalling? Maintenance of hearing in the asexual bushcricket *Poecilimon intermedius*. J Comp Physiol A 193:537–545
Lehmann GUC, Berger S, Strauß J, Lehmann AW, Pflüger HJ (2010) The auditory system of non-calling grasshoppers (Melanoplinae: Podismini) and the evolutionary regression of their tympanal ears. J Comp Physiol A 196:807–816
Legendre F, Robillard T, Song H, Whiting MF, Desutter-Grandcolas L (2010) One hundred years of instability in ensiferan relationships. Syst Entomol 35:475–488
Lewis DB (1974) The physiology of the tettigoniid ear. I. The implications of the anatomy of the ear to its function in sound reception. J Exp Biol 60:821–837
Meier T, Reichert H (1990) Embryonic development and evolutionary origin of the orthopteran auditory organs. J Neurobiol 21:592–610
Michel K (1979) Ein neuer Typ eines Grillenohres: Konstruktion und Ultrastruktur eines Tympanalorgans von *Cycloptiloides canariensis* (Bolivar) (Mogoplistinae, Gryllidae). Zool Anzeiger 203:139–150
Miller LA (1970) Structure of the green lacewing tympanal organ (*Chrysopa carnea*, Neuroptera). J Morphol 131:359–382
Moiseff A, Pollack GS, Hoy RR (1978) Steering responses of flying crickets to sound and ultrasound: mate attraction and predator avoidance. Proc Natl Acad Sci USA 75:4052–4056
Moulins M (1976) Ultrastructure of chordotonal organs. In: Mill PJ (ed) Structure and function of proprioceptors in the invertebrates. Chapman and Hall, London, pp 387–426
Nocke N (1975) Physical and physiological properties of the tettigoniid ('grasshopper') ear. J Comp Physiol 100:25–57
Otte D (1990) The relation between hearing and flying in crickets. Entomol News 101:29–34
Pearson KG, Hedwig B, Wolf H (1989) Are the hind wing chordotonal organs of the locust elements in the flight pattern generator? J Exp Biol 144:235–255
Plotnick RE, Smith DM (2012) Exceptionally preserved fossil insect ears from the Eocene Green River formation of Colorado. J Paleontol 86:19–24
Pollack GS (2000) Who what where? Recognition and localization of acoustic signals by insects. Curr Opin Neurobiol 10:763–767
Pollack GS, Martins R (2007) Flight and hearing: ultrasound sensitivity differs between flight-capable and flight-incapable morphs of a wing-dimorphic cricket species. J Exp Biol 210:3160–3164
Prager J (1976) Das mesothorakale Tympanalorgan von *Corixa punctata* Ill. (Heteroptera, Corixidae). J Comp Physiol 110:33–50
Radl E (1905) Über das Gehör der Insekten. Biol Centralblatt 25:1–5
Regier JC, Zwick A, Cummings MP, Kawahara AY, Cho S, Weller S, Roe A, Baixeras J, Brown JW, Parr C, Davis DR, Epstein M, Hallwachs W, Hausmann A, Janzen DH, Kitching IJ, Solis M, Yen S-H, Bazinet AL, Mitter C (2009) Towards reconstructing the evolution of advanced moths and butterflies (Lepidoptera: Ditrysia): an initial molecular study. BMC Evol Biol 9:280
Riede K (1987) A comparative study of mating behaviour in some neotropical grasshoppers (Acridoidea). Ethology 76:265–296
Rodriguez RL, Greenfield MD (2004) Behavioural context regulates dual function of ultrasonic hearing in lesser waxmoth: bat avoidance and pair formation. Physiol Entomol 29:159–168
Roeder KD (1967) Nerve cells and insect behavior. Harvard University Press, Cambridge
Rössler W (1992) Functional morphology and development of tibial organs in the legs I, II and III of the bushcricket *Ephippiger ephippiger* (Insecta, Ensifera). Zoomorphology 112:181–188

Rust J, Gottwald J, Stumpner A (1999) Singing and hearing in a Tertiary bushcricket. Nature 399:650

Schäffer S, Lakes-Harlan R (2001) Embryonic development of the central projection of auditory afferents (*Schistocerca gregaria*, Orthoptera, Insecta). J Neurobiol 46:97–112

Schumacher R (1979) Zur funktionellen Morphologie des auditiven Systems der Laubheuschrecken (Orthoptera: Tettigonioidea). Entomol Generalis 5:321–356

Shaw SR (1994) Re-evaluation of the absolute threshold and response mode of the most sensitive know vibration detector, the cockroach's subgenual organ: a cochlea-like displacement threshold and a direct response to sound. J Neurobiol 25:1167–1185

Simmons NB, Seymour KL, Habersetzer J, Gunnell GF (2008) Primitive early Eocene bat from Wyoming and the evolution of flight and echolocation. Nature 451:818–821

Stölting H, Stumpner A, Lakes-Harlan R (2007) Morphology and physiology of the prosternal chordotonal organ of the sarcophagid fly *Sarcophaga bullata* (Parker). J Insect Physiol 53:444–454

Strauß J, Lakes-Harlan R (2008a) Neuroanatomy and physiology of the complex tibial organ of an atympanate Ensiferan, *Ametrus tibialis* (Brunner von Wattenwyl 1888) (Gryllacrididae, Orthoptera) and evolutionary implications. Brain Behav Evol 71:167–180

Strauß J, Lakes-Harlan R (2008b) Neuroanatomy of the complex tibial organ of *Stenopelmatus* (Orthoptera: Ensifera: Stenopelmatidae). J Comp Neurol 511:81–91

Strauß J, Lakes-Harlan R (2009) The evolutionary origin of auditory receptors in Tettigonioidea: the complex tibial organ of Schizodactylidae. Naturwissenschaften 96:143–146

Strauß J, Lakes-Harlan R (2010) Neuroanatomy of the complex tibial organ of the splay-footed cricket, *Comicus calcaris* (Orthoptera: Ensifera: Schizodactylidae). J Comp Neurol 518:4567–4580

Strauß J, Lehmann GUC, Lehmann AW, Lakes-Harlan R (2012) Spatial organization of tettigoniid auditory receptors: insights from neuronal tracing. J Morphol 273:1280–1290

Stumpner A, Heller K-G (1992) Morphological and physiological differences of the auditory system in three related bushcrickets (Orthoptera: Phaneropteridae, *Poecilimon*). Physiol Entomol 17:73–80

Stumpner A, von Helversen D (2001) Evolution and function of auditory systems in insects. Naturwissenschaften 88:159–170

Surlykke A (1984) Hearing in notodontid moths: a tympanic organ with a single auditory neuron. J Exp Biol 113:323–335

van Staaden MJ, Römer H (1998) Evolutionary transition from stretch to hearing in ancient grasshoppers. Nature 394:773–776

Webster DB, Fay RR, Popper AN (1992) Comparative and evolutionary biology of hearing. Springer, New York

Weissman DB (2001) Communication and reproductive behaviour in North American Jerusalem crickets (*Stenopelmatus*) (Orthoptera: Stenopelmatidae). In: Field L (ed) The biology of wetas, king crickets and their allies. CABI Publishing, Wallingford, pp 351–373

Wheeler WC, Whiting M, Wheeler QD, Carpenter JM (2001) The phylogeny of the extant hexapod orders. Cladistics 17:113–169

Wyttenbach RA, May ML, Hoy RR (1996) Categorical perception of sound frequency by crickets. Science 273:1542–1544

Yack JE, Fullard JH (1990) The mechanoreceptive origin of insect tympanal organs: a comparative study of similar nerves in tympanate and atympanate moths. J Comp Neurol 300:523–534

Yack JE, Scudder GGE, Fullard JH (1999) Evolution of the metathoracic tympanal ear and its mesothoracic homologue in the Macrolepidoptera (Insecta). Zoomorphology 119:93–103

Yack JE, Fullard JH (2000) Ultrasonic hearing in nocturnal butterflies. Nature 403:265–266

Yack JE, Hoy RR (2003) Hearing. In: Resh VH, Carde RT (eds) Encyclopedia of insects. Academic Press, San Diego, pp 498–505

Yack JE (2004) The structure and function of auditory chordotonal organs in insects. Microsc Res Tech 63:315–337

Yager DD (1990) Sexual dimorphism of auditory function and structure in praying mantises Mantodea; Dictyoptera. J Zool 221:517–537
Yager DD (1999) Structure, development, and evolution of insect auditory systems. Microsc Res Tech 47:380–400
Yager DD (2005) Cockroach homologs of praying mantis peripheral auditory system components. J Morphol 265:120–139
Yager DD, Svenson GJ (2008) Patterns of praying mantis auditory system evolution based on morphological, molecular, neurophysiological, and behavioural data. Biol J Linn Soc 94:541–568
Yager DD (2012) Predator detection and evasion by flying insects. Curr Opin Neurobiol 22:201–207
Yager DD, Spangler HG (1995) Characterization of auditory afferents in the tiger beetle, *Cicindela marutha* Dow. J comp Physiol A 176:587–599

Chapter 3
Hearing and Sensory Ecology of Acoustic Communication in Bladder Grasshoppers

Heiner Römer, Adam R. Smith and Moira van Staaden

Abstract Bladder grasshoppers are a small family of Orthoptera, with ear morphology and physiology, behavior, and sensory ecological features outstanding among acoustic insects. Acoustic communication is characterized by male and female duetting and male phonotaxis. The detection distance of the male signal is exceptional at about 2 km, achieved via stridulation against air-filled abdominal resonators, and exploitation of weather conditions ideal for sound transmission. In at least three species, alternate male morphs occur which are incapable of flight and sound production but copulate with females. Such alternative mating tactics constitute profound selective pressures for sexual competition and the evolution of the communication system. Auditory sensitivity is mediated by an array of six pairs of atympanate ears in abdominal segments A1–A6. The auditory organ, a pleural chordotonal organ, in A1 comprises about 2,000 sensilla, whereas ears in segments A2–A6 are less developed, making pneumorids a unique system for studying the evolution of complex ears from simple precursors.

H. Römer (✉)
Institute for Zoology, Karl-Franzens-University, Universitätsplatz 2, 8010 Graz, Austria
e-mail: heinrich.roemer@uni-graz.at

A. R. Smith
Department of Biology, Indiana University, 1001 East 3rd St, Jordan Hall 142, Bloomington, IN 47405, USA
e-mail: adarsmit@indiana.edu

M. van Staaden
Department of Biological Sciences, Bowling Green State University, 303 Life Sciences Building, Bowling Green, OH 43403, USA
e-mail: mvs.bgsu@gmail.com

B. Hedwig (ed.), *Insect Hearing and Acoustic Communication*,
Animal Signals and Communication 1, DOI: 10.1007/978-3-642-40462-7_3,

3.1 Introduction

Putting into the port of Simon's Town on the evening of 31 May 1836, Charles Darwin claimed that he "saw so very little worth seeing, that I have scarcely anything to say."(pp 425–426 voyage diary; Darwin and Keynes 1988). Had he arrived at the Cape of Good Hope just 8 weeks later, his initial impressions of the southern African subcontinent may have been substantially different. Several decades after, Darwin was to examine a grasshopper in the British Museum collection and recognize it as profoundly modified for the sake of stridulation "…for in the male the whole body has been converted into a musical instrument, being distended with air, like a great pellucid bladder, so as to increase the resonance." (Darwin 1871).

These bladder grasshoppers (Orthoptera; Pneumoridae) comprise an ancient family of Acridid grasshoppers, endemic to the coastal regions of Africa (Dirsh 1965; Flook and Rowell 1997). In situ, Darwin would have heard the males making "a wonderful noise during the night" (p 359; 1871) and seen them gathering around lights in large number. Besides the exaggerated nocturnal signaling of macropterous males though, it is unlikely that he would have uncovered much more in his brief visit, for the 17 currently recognized pneumorid species are highly cryptic, host plant specificity is strong, and all other micropterous individuals are confined to their food plants. A richly complex environment, much phenological specialization, and a plant diversity making it one of the most species-rich areas on earth (Linder 2003), combine to disperse bladder grasshoppers patchily, and at low densities, across space and time. Pneumorids thus face the challenge of navigating social interactions and locating mates under particularly demanding circumstances and it is unsurprising then, that their most notable morphological and behavioral features are those relating to acoustic communication.

A comparison of acoustic behavior and the related sensory system in this group with those of modern, short-horned grasshoppers reveals some striking similarities and differences. For example, in one of the best studied species of Acridid grasshopper, *Chorthippus biguttulus*, pair formation and mating is initiated by an acoustic duet in which males produce a song with species-specific amplitude modulation, and receptive females respond with a song used by the male to orient toward her (see Chap. 10 by Ronacher). Similarly, it is the male in bladder grasshoppers which initiates pair formation by producing a loud, stereotyped calling song at night. Males call at irregular intervals from a stationary position high up in the vegetation, and they may move distances up to 500 m between successive calls if they do not receive a female reply. Receptive females within hearing range for the male call may respond with a low-intensity call within a fixed time window 720–860 ms after the end of male call. The female response then induces a duet and male phonotaxis, where he makes a short flight, moves up to a high point on the vegetation before calling again, reorienting and repeating the procedure until finally contacting the female. Remarkably, the orientation of the male is extremely direct and accurate once he is within the hearing range of a female response. Given that adjustments must be made in both azimuth and elevation, this performance is reminiscent of the excellent orientation of the parasitoid fly *Ormia ochracea* when approaching its singing male cricket host (Müller and Robert 2001).

In contrast to Gomphocerine grasshoppers, where duetting takes place over only 1–2 m, the distance covered by bladder grasshoppers is in the range of 100 m, and hearing distance can be close to 2 km (van Staaden and Römer 1997). There are three factors contributing to the large discrepancy in the communication range between modern and bladder grasshoppers: (1) differences in call amplitude and spectrum, (2) ideal transmission of the calls under nocturnal conditions, and (3) high sensitivity of hearing organs.

This chapter addresses how bladder grasshoppers detect sounds, decipher meaning in auditory information, and use this to direct adaptive behavior. We focus first on the exaggerated acoustic signaling of duetting pairs, reviewing internal and external filtering imposed by the nervous system and the environmental transmission channel, respectively. We then consider the evolutionary impacts and constraints levied by the presence of a broader audience and finally, entertain the possibilities inherent in a perceptual allocation approach to pneumorid communication.

3.2 Combining Bladders and Acoustic Adaptation Facilitates Record-Breaking Communication Distances

3.2.1 Sound Production Using Resonators

The calling song of males is produced when a scraper with a small row of strong, transverse ridges on the proximal side of the hind femur is moved against a file of strongly sclerotized ridges on the second abdominal tergite. The abdomen of adult males is characterized by a permanently inflated bladder, giving the name to the whole family of bladder grasshoppers. The air-filled abdominal cavity acts as a resonator when the impact of file and scraper is spread across its large surface, creating a sound output of 98 dB SPL at 1 m in the best studied species *Bullacris membracioides* (van Staaden and Römer 1997). In this species, the male call consists of five short, 'noisy' syllables and a sixth long, resonant syllable with its main energy centered around 1.7 kHz, which is unusually low for acridid orthoptera. In other species, the range of carrier frequencies is between 1.5 and 3.2 kHz (Couldridge and van Staaden 2004). The SPL of the short introductory syllables is reduced by 20–25 dB relative to the final syllable.

There is a clear sexual dimorphism in the signals used for acoustic duetting, both in the mechanism by which they are produced and in the acoustic structure and loudness. The female response is rather soft compared to the final syllable of the male call (60 dB SPL at 1 m), and is produced by rubbing teeth-bearing veins on the ventral margins of the wings across raised pegs in a differentiated region of the tergum beneath the resting wing. The female call is produced in series of 1–8 syllables (depending on the perceived SPL of the male call; see below), with a frequency spectrum from 3 to 11 kHz.

Altogether, the characteristics of male and female calls in the acoustic duet, and the roles of both sexes in pair formation are rather typical for duetting species (see review by Bailey 2003 for similar cases in duetting *Phaneropterinae katydids*): it is the male covering most of the costs associated with conspicuous signaling, as well as locomotion toward the female (Zuk and Kolluru 1998). These high costs of signaling and movement might have been important for the evolution of alternate male forms and mating tactics. Furthermore, since the signal with the smallest active range limits acoustic communication in duetting species (Zimmermann et al. 1989), it is the soft female reply and not the exaggerated male call which finally determines the range over which pair formation in bladder grasshoppers takes place.

3.2.2 Sound Transmission in the Natural Habitat

Although it is quite uncertain whether maximum range of detection is the primary selection pressure on animal vocalizations (Michelsen 1978; Richards and Wiley 1980; Römer 1998; Wiley and Richards 1978, 1982), bladder grasshoppers provide one of the best examples of how animals might use constraints imposed by the acoustic conditions of the habitat to maximize broadcast range of their signals. Observations of calling times and meteorological conditions at night revealed that acoustic communication only occurred after strong temperature inversions formed at the surface shortly after sunset, accompanied by calm wind conditions with speeds of <2 m/s. This contrasts with the super-adiabatic conditions from mid-morning to mid-afternoon, in which temperature drops rapidly with height above the hot ground surface. Strong differences were observed when signal transmission through the natural environment of the insect was determined for these two atmospheric conditions (van Staaden and Römer 1997). During the afternoon, the super-adiabatic situation produced an upward refracting of sound and a sound shadow zone, with a consequent marked drop and high variability in SPL of the male signal starting at a distance of about 50 m. By contrast, temperature inversions after sunset were downward refracting, resulting in a tunnel effect where the sound was caught between these zones of different temperature and the ground. Therefore, attenuation of the male call approached ideal values according to geometrical spreading of sound for distances up to 450 m at a nocturnal time when males and females actually communicate. Due to these rather different atmospheric conditions, hearing distances for the male signal are 120–200 m in the afternoon, but between 1.4 and 1.9 km at night, arguably the largest hearing distance yet reported for insects.

Since pneumorids' success in mate detection is entirely dependent on long-distance acoustic signaling, one may reason that natural selection should have molded signal form and transmission to maximize communication efficacy. Moreover, across the entirety of their geographical distribution range, the pneumoroidea occupy diverse biomes from open savanna and succulent karoo, averaging 0.5 m in vegetation height, to the more complex heathland vegetation ("fynbos"), and forest biomes topping out at 20 m. This combination of features provides an ideal test case for the Acoustic Adaptation Hypothesis (AAH), which predicts that the transmission efficiency of long-distance signals should be greater in native than non-native habitats (Morton 1975). The performance of male advertisement calls of bladder grasshoppers were directly compared in forest, fynbos, savanna, and succulent karoo habitats (Couldridge and van Staaden 2004). Transmission distance and signal fidelity measures indicated that the physical structure of forest and fynbos biomes imposes strong selection pressure on signal production of their native species. Additional challenges in these two habitats are posed by the highest levels of ambient/biotic noise in forests, and inconsistency in transmission efficacy/signal degradation in fynbos. Together, these selective forces have molded communication signals in forest and fynbos species which propagate with lower levels of distortion over distance in their native habitats, and in the case of fynbos taxa, have lower levels of signal attenuation over distance in all but the forested environment. Many of the design features for superior (distance/fidelity) transmission predictably accord with predictions of the AAH, e.g., repetitive syllable elements in dense habitats, and short signals with rapid rise time in open ones, though *B. obliqua* native to fynbos has a very gradual rise time over signals of 6 s duration. Environment-related adjustments in the structure of vocal signals appear to be constrained by additional selective forces in the succulent karoo and savannah biomes, and there is thus no overall support for AAH across all pneumoridae. This is clearly not due to constraints by call function as frequently found in anurans, mammals, and cricket assemblages (Ey and Fischer 2009; Jain and Balakrishnan 2012). However, the test is a conservative one, with transmission experiments conducted under rather narrowly defined wind/temperature conditions, and it is unclear whether the results would hold up under more variable weather conditions.

3.3 Serial, Non-tympanal Hearing Organs Provide the Sensory Basis for Sophisticated Behavior

Given the detailed duetting behavior and the sophisticated use of atmospheric conditions at night for maximizing the active range of the signal, it is a surprise to note the absence of tympana in the abdominal hearing organs, which are typical for ears of modern grasshoppers. In order to determine the neural basis underlying pair formation behavior in *B. membracioides,* van Staaden and Römer (1998) and van Staaden et al. (2003) used anatomical, neurophysiological, and behavioral analyses to define the extent to which this species possesses functional ears despite the lack of tympana.

At the same position, laterally in the first abdominal segment where in modern grasshoppers a pair of tympanal ears is located, the bladder grasshopper lacks a thinned tympanic membrane, but nevertheless, there is a pear-shaped pleural

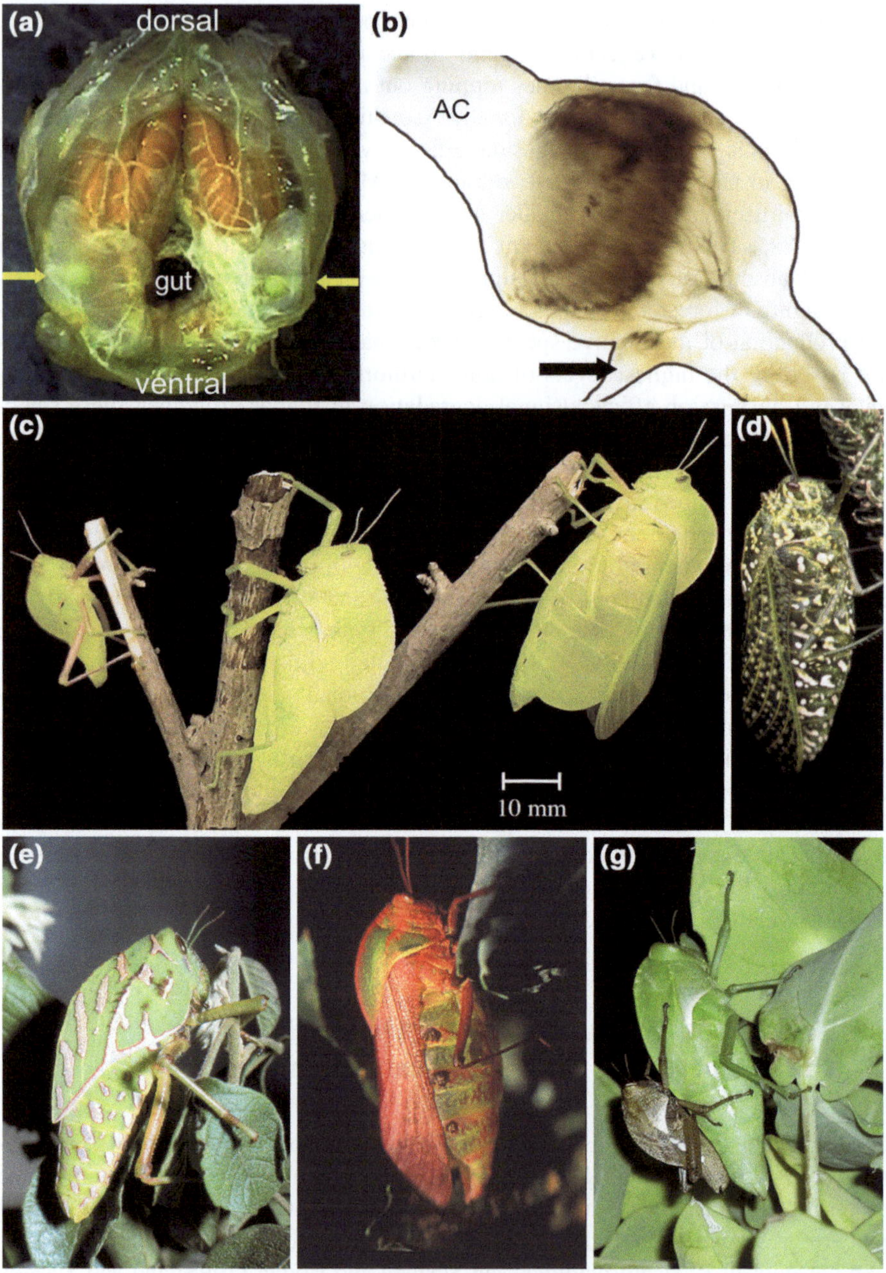

chordotonal organ (plCO1) in a corresponding location as the Müller's organ (Fig. 3.1a). The receptor organ in Bullacris is much larger in size and houses about 2,000 sensilla, compared to only 80 in the locust (Gray 1960; Jacobs et al. 1999). A striking similarity between plCO1 and Müller's organ is the attachment to the

◀ **Fig. 3.1** **a**, **b** Location and structure of the pleural chordotonal organ (plCO1) in abdominal segment 1. **a** Dissection of the A1 segment, a pair of pleural chordotonal organs is located in air-filled cavities (*arrows*), where the attachment cells of the scolopidia attach directly to the cuticle of the abdomen, without tympanal specializations. **b** Retrograde cobalt backfill of plCO1 *B. membracioides* through the sensory nerve labels axons, cell bodies, and dendrites of the sensory cells. Note one small group of sensory cells (*arrow*) with a connection to the cuticle via a separate thin bundle of attachment cells. **c**–**g** Bladder grasshopper species and morphs. **c** From left to right: adult forms of *B. membracioides* uninflated, alternative male, female, inflated male. **d** Adult male *Physemacris variolosus*. **e** Female *Pneumora inanis*, **f** adult male *B. discolor*, **g** alternate male and adult female *B. unicolor*

pleural cuticle of A1 via a thick and a thin bundle of very long attachment cells separated by more than 1 mm (Fig. 3.1b). In Müller's organ as well, one group of sensory cells (d-cells) connects to the tympanum via a thin bundle of attachment cells to a site different from the majority of receptor cells. Only about 30 sensilla in Bullacris have their attachment cells within the thin bundle. This morphological arrangement suggests that this small group of sensilla may well represent the ancestral precursors of the 12–14 high-frequency d-cells in modern grasshoppers, although the frequency tuning of these sensilla is currently unknown. Each sensillum comprises a bipolar sensory cell, a scolopale cell, an attachment cell, and a glial (Schwann) cell. At the ultrastructural level, sensilla in Bullacris conform to the basic structure of scolopidial sensilla as in other insects and mechanoreceptive organs (Yack and Fullard 1993). In Bullacris, five further pairs of plCOs exist in abdominal segments A2–A6, suspended between the sternal apodeme on one side and their site of attachment to the lateral body wall on the other. The plCOs 2–6 contain up to 11 sensilla per organ.

Extracellular multiunit recordings from the afferent nerves carrying the axons of pleural organs in A1–A6 revealed their nature as functional, serial homologous ears. All receptors responded to acoustic stimulation within a biologically meaningful intensity and frequency range, but with differences in tuning and thresholds. The best frequency of plCO1 was 4 kHz, and thus mismatched to the male song's carrier frequency of 1.7–2 kHz. Surprisingly, despite the absence of an overt tympanum, the organ is extremely sensitive at its best frequency with an average threshold of about 13 dB SPL in normal males, and about 20 dB SPL in alternate males and females. By contrast, receptors in plCOs 2–6 had best frequencies that matched the carrier frequency of the conspecific male signal (Fig. 3.2), but were significantly less sensitive with thresholds ranging from about 60–75 dB SPL.

The physiological background for the tuning of the pleural organs in the segments A2–A6 "matched" to the male call is currently unknown, but cannot be attributed to the mechanical resonance properties of the inflated abdomen since (1) the same tuning is present also in females lacking the inflated abdomen and (2) male and female pleural organs do not differ significantly in their thresholds. By comparing a number of atympanal chordotonal organs in other insects, we suggested that tuning to mechanical oscillations close to 2 kHz is a common property of such organs rather than an adaptation to the carrier frequency of the male call in *B. membracioides*. The list includes chordotonal organs on various body parts

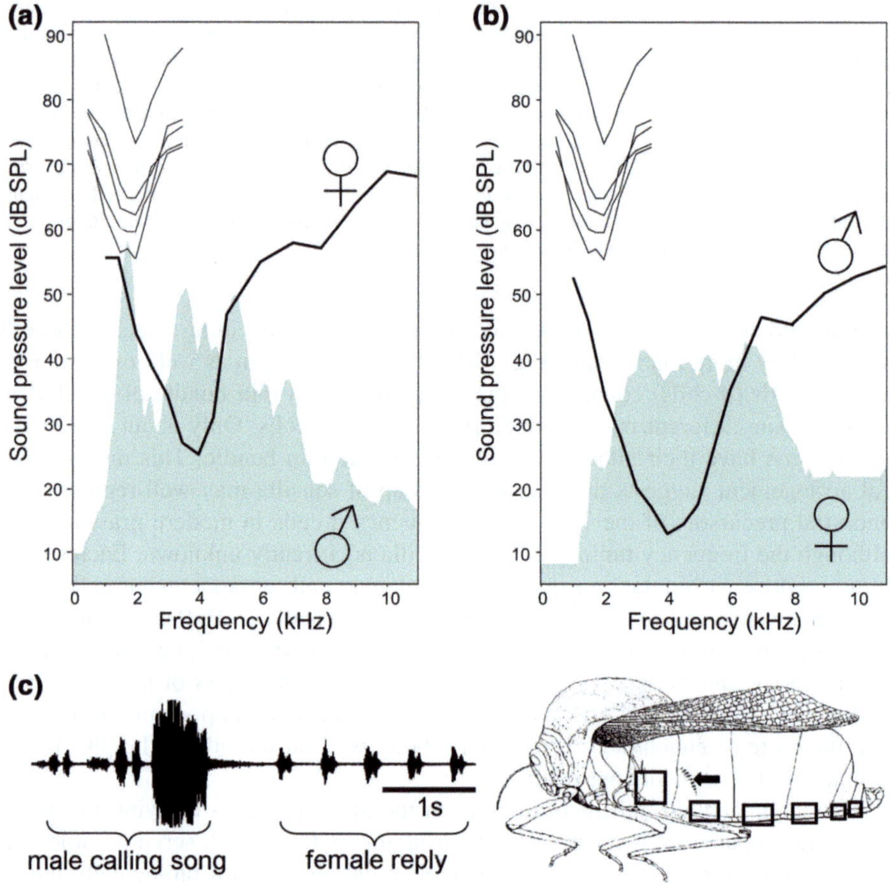

Fig. 3.2 Power spectra of male–female duetting signals (*shaded areas*) and neurophysiological tuning curves in *B. membracioides* for plCO1 (*bold line*) and plCO of A2–A6 (fine lines, not to scale) for a receiver of the opposite sex. **a** Male sender, female receiver; **b** female sender, male receiver ($n = 10$ individuals). **c** Adult male with strongly inflated abdomen. Location of stridulatory file arrowed; location of plCOs in A1–A6 indicated by rectangles. The oscillogram shows a male–female duet in *B. membracioides*

in different taxa (Pflüger and Field 1999; Yack and Fullard 1990; Cokl et al. 1995; Shaw 1994; Cokl and Virant-Doberlet 1997; Yager 1990). The organs exhibit also variable thresholds from 55 to 80 dB SPL, similar to the range of thresholds covered by the plCOs in segments A2–A6 of the ancestral insect studied here. This suggests that mechanical tuning is unlikely and favors an explanation intrinsic to the sensory neurons, as discussed for the tuned afferents in the crista acustica of katydids (Oldfield 1985). If this is the ancestral condition of chordotonal organs for the detection of airborne sound in acridids or pneumorids, it would represent a preadaptation for the evolution of a long-distance male call at a frequency where the female organs are most sensitive (see Chap. 2 by Strauß and Lakes-Harlan).

3.4 The Functional Significance of Pleural Hearing Organs in Male–Female Duets

The data in the previous sections demonstrate, on the one hand, a sexual dimorphism in the signals used in the acoustic duet between male and female bladder grasshoppers, and on the other hand a difference in tuning and sensitivity in the serial homologous ears in A1 compared to A2–A6. As indicated in Fig. 3.2, plCO1-afferents are not tuned to the main CF of the male call, but would be most sensitive to components of the female reply. Indeed, in behavioral playback experiments with receptive females using model songs of males with carrier frequencies either at 1.7 or 4 kHz, almost no female reply was elicited with the higher frequency, but close to 100 % with the lower one. Since the playback intensity was 75 dB SPL for both frequencies, the plCO1 was stimulated more than 40 dB above its threshold, in contrast to only 15 dB for the pleural organs in A2–A6. Even after the ablation of plCO1-receptors on both sides, the behavioral response of females was unaffected, indicating that their reply in duetting is mediated on the afferent side only by these less sensitive pleural organs (van Staaden and Römer 1998). A further indication for the direct role of the pleural ears in A2–A6 comes from the observation that females add approximately one syllable to their acoustic reply with each 3 dB increase in SPL of the male song above the threshold of the behavioral response at about 60 dB SPL, which is also the threshold of the pleural organ in A2. Apart from this strong correlation between activation of more pleural organs with increasing SPL of the male call and the increase in the behavioral response, this system is also remarkable in that the female, via the number of syllables in her reply, appears to signal to the male very reliably her proximity.

For males it is tempting to speculate, based on the overlap of call energy and tuning of plCO-sensory neurons in A1, that their function is the detection and discrimination of the much softer female call. In this context, it would indeed be rather adaptive for males to evolve the high sensitivity as observed in the physiological responses, in order to detect a responding female at the greatest possible distance. However, in order to demonstrate a causal relationship between the plCO1 activity and the acoustic duetting and possible phonotactic behavior of males, this would require a similar ablation experiment as in females, but this time the ablation of all five pairs of pleural organs in A2–A6, which has not been done so far.

3.5 Frequency and Intensity Discrimination may Provide Distance Estimation

Members of the Pneumoridae are the only known insects with serially repeated, functional ears which differ in their tuning (if we exclude those taxa with tympanate hearing and additional sensitivity to low frequency sound of wind-sensitive hair sensilla on the cerci). So far we do not know whether the 2,000 receptors in plCO1 are tuned to the same or different frequencies. But given the homology in structure with the Müller's organ of modern grasshoppers, the few receptors with their attachment site far away from the main bundle could be tuned to higher frequencies, as the d-cells in locusts and other grasshoppers (Michelsen 1971; Römer 1976; Jacobs et al. 1999). The fact that females do not respond to song models with a carrier of 4 kHz (the best frequency of plCO1-receptors) indicates some kind of frequency discrimination, although this is not as elaborated as the categorical perception of sound frequency in crickets (Wyttenbach et al. 1996; Moiseff et al. 1978), or the discrimination of the sexes by spectral differences in the signals in some grasshoppers and katydids (Dobler et al. 1994; von Helversen and von Helversen 1997).

Probably, more important than the tuning of receptors and corresponding frequency discrimination is the evaluation of actual sender–receiver distances during duetting. Females modify their reply with distance from the calling male, adding about one syllable per 3 dB increase in loudness (van Staaden and Römer 1998). The sensory basis for this behavior could be the general intensity-response characteristics of receptors, firing at a higher rate with increasing stimulus amplitude. The fact that in Bullacris all plCOs in A2–A6 are tuned to the same frequency of about 2 kHz, but differ in their absolute thresholds (Fig. 3.2) offers a more likely solution (see Fig. 3.3): depending on distance and thus loudness of the male call, more and more hearing organs are stimulated, and within the population of 11 afferents in each organ, the firing increases as well. Thus at a distance of 30 m plCOs in A5 and A6 do not respond at all to the male call, whereas those in A4–A2 respond only to the last, final syllable (Fig. 3.3). Closer to 8 m, the final syllable is suprathreshold even for the least sensitive A6 receptors, and in addition even the soft introductory syllables of the male call are faithfully encoded in the discharge of neurons in A2 and A3. Thus, the number of pleural organs activated, and the degree of activation within each organ provides the female with reliable information about distance to the male. Such a coding scheme, where the number of receptor neurons being suprathreshold increases with increasing stimulus amplitude is known as range fractionation (for review see Hedwig and Pollack 2008). But in contrast to other insect species where more afferents in a single organ are recruited for encoding the distance to the signaler (Roeder and Treat 1957; Römer

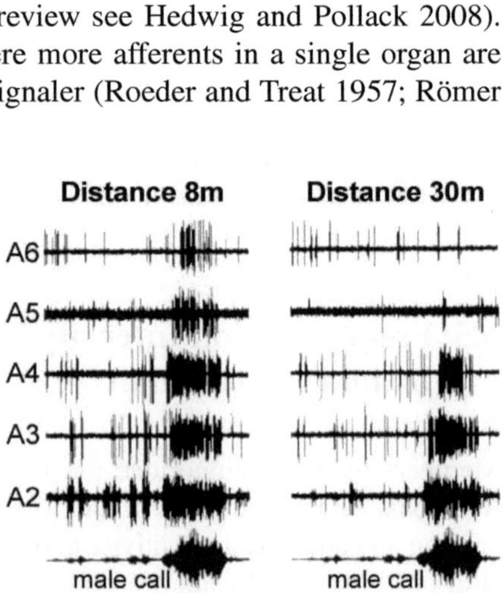

Fig. 3.3 Neural responses of pleural hearing organs and distance perception. Extracellular multiunit recording of the activity of female plCO in A2–A6 in response to a male call perceived at distances of 8 m and 30 m

1987; Oshinski and Hoy 2002; Römer et al. 1998), in Bullacris additional serial repeated ears are recruited for this task.

Interestingly, observations in the field indicate that the SPL of the male call is not always close to 100 dB, in particular when the male has established reliable duetting with the female, when both are within the active range of their signals. In these cases, males often down-regulate the SPL of their call, with the consequence that the above-described activation of pleural receptors in the female decreases again. However, the decrease in loudness of the male call happens at close range (distance about 10 m) when all pleural organs are activated, and apparently is not strong enough to interrupt the duet. One selection pressure for this behavior might be the competition by "unintended receivers" from conspecific males, particularly alternate male morphs (Zuk and Kolluru 1998; McGregor 2005).

3.6 A Diversity of Signals and Receivers: What is Being Signaled and to Whom?

The nature of the duetting and the similarity of all female responses across pneumorid taxa indicate that females are responsible for the recognition/choice continuum in an intersexual context. However, short-term adjustments of male signals to social conditions allude to the presence of additional selection pressures. In all cases though, the challenges of communicating at a distance requires a bladder grasshopper to (i) detect signals against background noise, (ii) identify signals as stemming from a conspecific, and (iii) rate the attractiveness of signals.

Long-range signaling results in significant call degradation along the transmission channel, and the expectation is that signals will become less distinguishable and therefore less attractive at greater distances. Playback of degraded male calls in the absence of intensity cues indicated that both signal quality and amplitude are important determinants of preference for pneumorid females. Although females responded to signals with degradation levels equivalent to a male calling 150 m away, but intensity equivalent to one at 25 m, suggesting that call amplitude is a limiting factor, responses to conspecific calls decline significantly with decreased signal quality (Couldridge and van Staaden 2006). Unlike *Ch. biguttulus* (Einhäupl et al. 2011), pneumorid males have not solved the problem of retaining attractiveness cues in the face of signal degradation. Specific identity, on the other hand, is contained in the male call of at least some taxa. In interspecific preference tests with sympatric congeners *B. intermedia* and *B. serrata, B. membracioides* females discriminate only against the calls of *B. serrata* (Couldridge and van Staaden 2006). They also exhibit preferences for the songs of particular conspecific males in a typical sexual selection scenario where males provide females no resources or direct benefits (Couldridge and van Staaden 2006). However, it is difficult to identify the precise call features that are responsible for these differences in attractiveness. In playbacks to female *B. membracioides*, multiple regressions identified all but one of eight measured call features (intersyllable interval) as significantly correlated with female preference. Enhanced female responses appear to be driven by short/high-frequency introductory syllables, and longer/low frequency final syllables, it would be interesting to know whether this is a step function, and at what combination of features (i.e., duration, frequency, or rise time) the differential female response saturates.

3.7 Alternate Males

In at least three species of pneumorid, there are alternate males which are incapable of flight and sound production because they lack macropterous wings and the inflated abdomen, but retain the strong host-plant philopatry of nymphal stages and can be found in the field in copula with females (Alexander and van Staaden 1989; Fig. 3.1). Despite the fact that inflated and alternate morphs follow distinctly different developmental trajectories in attaining their final forms (Donelson and van Staaden 2005), tuning and sensitivity of their hearing organs is identical (van Staaden et al. 2003). Costs and benefits for polyphenic males differ strongly though, with adult longevity of small, uninflated males twice that of the larger inflated ones (Donelson et al. 2008). Both morphs exhibit positive phonotaxis to conspecific female calls in playback experiments, but differ markedly in their response to male signals. In this situation, primary males move perpendicular to the stimulus source, whereas alternates remain stationary (Donelson and van Staaden 2005). Rather than actively searching for mates over great distances then, alternate males eavesdrop on duets and intercept responsive females before the normal, flighted male arrives. Such satellite tactics exacerbate the already intense intrasexual selection posed by calling males.

The sensory and communication system is subjected to antagonistic selective forces emerging from the production of shorter range acoustic signals including female responses (ca. 50 m), disturbance signals (ca. 5 m), and putative territoriality signals emanating from the mouth and functional at very short range (<0.1 m) as well.

3.8 Acoustic Energy Catch and Perceptual Allocation

The complexity of pneumorid acoustic communication in terms of sensory infrastructure, signal range and transmission characteristics, repertoire size and potential audience raises many questions, which might best be characterized as interaction effects. Perceptual allocation approaches attempt to mimic the basic features of sensory systems, and have been used to explore the key mechanisms underlying sensory detection and recognition as well as the hidden preferences that may emerge as a byproduct of such mechanisms (Enquist and Arak 1993; Phelps et al. 2001). As such, they may provide a useful complement to sensory ecology for understanding how the contingencies of the evolutionary process shape nervous systems to accommodate conflicting demands.

We find intriguing the suggestion by Phelps (2007) of an acoustic equivalent for the quantum catch concept, which has been exceptionally useful to those in the quantitative visual sciences (Anderson and Laughlin 2000; Stavenga 2004). In visual systems, the perception of chromatic signals can be approximated using quantum catch models that predict the absolute or relative amount of available light a sensory system can "catch," and this is done regularly, e.g., in aquatic visual ecology (reviewed briefly by Smith et al. 2012). However, it is important to note that the general principles underlying quantum catch models can be extrapolated to the acoustic realm via the replacement of analogous variables (Phelps 2007) to estimate the sound energy captured by a receiver. As such, we can use the following equation to investigate auditory systems (Eq. 3.1):

$$AC_{abs,S} = \int_x^y S(f)\, T_e(f,d) R(f)\, df \tag{3.1}$$

Where $AC_{abs,S}$ is the absolute energy transduced from a given signal (henceforth called acoustic catch), x and y are the bounding frequencies for the signal calculation, $S(f)$ is the energy available in the signal at a given frequency, $T_e(f, d)$ is the transmission of the signal through the environment at distance d, and $R(f)$ is the neural sensitivity for the acoustic signal. This equation can be modified to yield the relative acoustic catch of the auditory system for any given signal (Eq. 3.2):

$$AC_{rel,S} = \frac{AC_{abs,S}}{\int_x^y S(f)\, T_e(f,d)} \tag{3.2}$$

Using these equations, we can effectively predict the ability of an acoustic organ to transduce the energy of any sound signal with four pieces of information: (i) frequency tuning curves for receptors in the organ, (ii) spectral characteristics of the signal in question, (iii) frequency-specific attenuation measures for the transmission through the environment, and (iv) background noise measures. In the pneumorids, we have one of the few insect systems where information is available on all of these elements.

Although for most insects it is the temporal structure of calls that is critical, we argue that acoustic catch is a reasonable approach for pneumorids because (1) of the importance of frequency and intensity discrimination (demonstrated in 3.6 above), (2) the frequency shifts observed when two calling males interact acoustically (unpublished data), and (3) due to the extent that the female acoustic response depends on a match between ear tuning and stimulus frequency (van Staaden and Römer 1998). Here, we make a first pass in simplified form using *B. membracioides* tuning curves obtained by recording the auditory nerve, male signal characteristics, and signal transmission (van Staaden and Römer 1998) without frequency-specific attenuation information, i.e., assuming the shape of the signal is unchanged with distance, which it most certainly is not. Some pneumorid taxa share overlapping geographic distribution throughout their range, so signals from heterospecifics can form a substantial part of their natural acoustic environment. To explore the power of the AC approach, we thus compared the relative

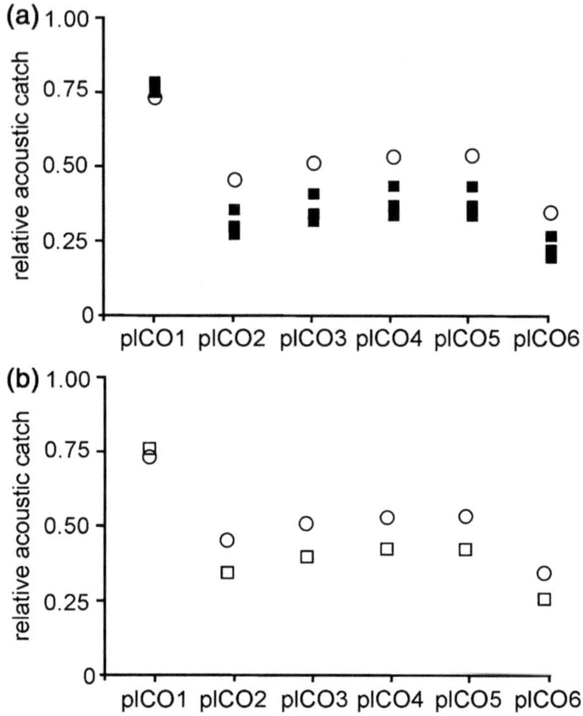

Fig. 3.4 **a** Relative acoustic catch (AC) of *B. membracioides* pleural hearing organs in A1–6 for conspecific (*open circles*) and heterospecific (*filled squares*) male calls. Relative acoustic catch represents the total acoustic catch of each plCO normalized to the total acoustic energy of the call for which the AC was calculated. Heterospecifics include taxa native to savanna (*B. intermedia, B. serrata*), forest (*Pneumora inanis*), fynbos (*B. obliqua, Physemacris variolosus*), and succulent karoo (*B. unicolor*). **b** Comparison of AC estimates for the auditory system of *B. membracioides* and a conspecific call (*open circles*) or a conspecific call shifted by 1.5 kHz toward higher frequencies (*open squares*)

stimulation of *B. membracioides* hearing organs for a conspecific male signal, as well as for signals of six heterospecifics.

For *B. membracioides*, plCOs catch proportions of available signal energy in the male call ranging from a high of 72 % in plCO1, through 44 % plCO2; 50 % plCO3; 52 % plCO4; 52 % plCO5, to 33 % in plCO6. Given the critical importance of acoustic signaling in mate detection and acquisition, it is also instructive to explore the catch of *B. membracioides* for heterospecific signals namely *B. intermedia, B. unicolor, B. serrata, B. obliqua, Physemacris variolosus,* and *Pneumora inanis*. Interestingly, whereas the plCO1 is equally sensitive to the calls (70–75 % "catch") of all these taxa, plCO2–6 perform better for the conspecific call than for any of the other taxa (Fig. 3.4). This is somewhat surprising given the relatively large overlap in frequency spectra of the different species' calls (see Figs. 3.1 and 3.2 in Couldridge and van Staaden 2004; 2006, respectively). Thus whereas *B. membracioides* hear female calls of all taxa equally well with the high sensitivity plCO1 receptors, a subset of hearing organs (plCO2–6) appear to be selective for conspecific signals. To explore how changes in signal characteristics might affect catch of the auditory organs, we shifted the sample *B. membracioides* call 1.5 kHz toward higher frequencies and found that the response of the auditory system decreased such that it closely matched the response to heterospecific signals (Fig. 3.4).

As an initial validation of the acoustic catch approach, these preliminary analyses demonstrate considerable potential, and suggest that the boundaries of signal plasticity within species may be rather narrowly delineated in the frequency domain. Combined with measures of frequency-dependent attenuation and tuning curves from additional taxa, such analyses could be used to test hypotheses concerning signal evolution, and both internal (sensory) and external (environmental) sources of filtering (for cases in crickets see Schmidt et al. 2011; Schmidt and Römer 2011).

Acknowledgments We thank the students and colleagues who collaborated on experiments in our laboratories and in the field. V. Couldridge and N. Donelson were responsible for the images of B. unicolor and *B. membracioides,* respectively. Research on which this chapter was based was supported by grants from the Austrian Science Foundation PO9523-BIO to HR, and the National Science Foundation 0091189 to MvS. Preparation of the chapter was supported by grant DUE 0757001 to MvS.

References

Alexander AJ, van Staaden MJ (1989) Alternative sexual tactics in male bladder grasshoppers (Orthoptera, Pneumoridae). In: Bruton MN (ed) Alternative Life-History Styles of Animals. Kluwer Academic Publishers, Dordrecht, pp 261–277

Anderson JC, Laughlin SB (2000) Photoreceptor performance and the co-ordination of achromatic and chromatic inputs in the fly visual system. Vision Res 40:13–31

Bailey WJ (2003) Insect duets: underlying mechanisms and their evolution. Physiol Entomol 28:157–174

Cokl A, Virant-Doberlet M (1997) Tuning of tibial organ receptor cells in *Periplaneta americana* L. J Exp Zool 278:395–404

Cokl A, Kalmring K, Rössler W (1995) Physiology of atympanate tibial organs in forelegs and midlegs of the cave-living Ensifera, *Troglophilus neglectus* (Raphidophoridae, Gryllacridoidea). J Exp Zool 273:376–388

Couldridge VCK, van Staaden MJ (2004) Habitat-dependent transmission of male advertisement calls in bladder grasshoppers (Orthoptera; Pneumoridae). J Exp Biol 207:2777–2786

Couldridge VCK, van Staaden MJ (2006) Female preferences for male calling songs in the bladder grasshopper *B. membracioides*. Behaviour 143:1439–1456

Darwin C, Keynes RD (eds) (1988) Charles Darwin's beagle diary. Cambridge University Press, Cambridge

Darwin C (1871) The descent of man, and selection in relation to sex. John Murray, London

Dirsh VM (1965) Revision of the family pneumoridae (Orthoptera: Acridoidea). Bull B M (N H) Ent 15:325–396

Dobler S, Stumpner A, Heller KG (1994) Sex-specific spectral tuning for the partner's song in the duetting bushcricket *Ancistrura nigrovittata* (Orthoptera: Phaneropteridae). J Comp Physiol 175:303–310

Donelson NC, van Staaden MJ (2005) Alternate Strategies in male bladder grasshoppers *Bullacris membracioides* (Orthoptera: Pneumoridae). Behaviour 142:761–778

Donelson NC, Smith AR, van Staaden MJ (2008) Variation in adult longevity in a polymorphic species. J Orthopt Res 17(2):279–282

Einhäupl A, Stange N, Hennig M, Ronacher B (2011) Attractiveness of grasshopper songs correlates with their robustness against noise. Behav Ecol. doi:10.1093/beheco/arr064

Enquist M, Arak A (1993) Selection of exaggerated male traits by female aesthetic senses. Nature 361:446–448

Ey E, Fischer J (2009) The "acoustic adaptation hypothesis"—a review of the evidence from birds, anurans and mammals. Bioacoustics 19:21–48

Flook PK, Rowell CHF (1997) The phylogeny of the Caelifera (Insecta, Orthoptera) as deduced from mtrRNA gene sequences. Mol Phylo Evol 8:89–103

Gray EG (1960) The fine structure of the insect ear. Phil Trans B 243:75–94

Hedwig B, Pollack GS (2008) Invertebrate auditory pathways. In: Basbaum AI, Karako A, Shepherd GM and Westheimer G (eds.) The senses: a comprehensive reference, Academic Press, San Diego, pp 525–564

von Helversen D, von Helversen O (1997) Recognition of sex in the acoustic communication of the grasshopper *Chorthippus biguttulus* (Orthoptera; Acrididae). J Comp Physiol A180:373–386

Jacobs K, Otte B, Lakes-Harlan R (1999) Tympanal receptor cells of *Schistocera gregaria*: correlation of soma positions and dendrite attachment sites, central projections and physiologies. J Exp Zool 283:270–285

Jain M, Balakrishnan R (2012) Does acoustic adaptation drive vertical stratification? A test in a tropical cricket assemblage. Behav Ecol 23:343–354

Linder HP (2003) The radiation of the Cape flora, southern Africa. Biol Rev 78:597–638

McGregor P (2005) Animal communication networks. Cambridge University Press, Cambridge

Michelsen A (1971) The physiology of the locust ear. J Comp Physiol A 71:49–126

Michelsen A (1978) Sound reception in different environments. In: Ali MA (ed) Sensory ecology. Plenum Press, New York, pp 345–373

Moiseff A, Pollack GS, Hoy RR (1978) Steering response of flying crickets to sound and ultrasound: mate attraction and predator avoidance. Proc Natl Acad Sci USA 75:4052–4056

Morton ES (1975) Ecological sources of selection on avian sounds. Am Nat 109:17–34

Müller P, Robert D (2001) A shot in the dark: the silent quest of a free-flying phonotactic fly. J Exp Biol 204:1039–1052

Oldfield BP (1985) The role of the tympanal membrane and the receptor array in the tuning of auditory receptors in bushcrickets. In: Kalmring K, Elsner N (eds.) Acoustic and vibrational communication in insects, Parey, pp. 17–24

Oshinski ML, Hoy RR (2002) Physiology of the auditory afferents in an acoustic parasitoid fly. J Neurosci 15:7254–7263

Pflüger HJ, Field LH (1999) A locust chordotonal organ coding for proprioceptive and acoustic stimuli. J Comp Physiol 184:169–183

Phelps SM (2007) Sensory ecology and perceptual allocation: new prospects for neural networks. Phil Trans R Soc B 362:355–367

Phelps SM, Ryan MJ, Rand AS (2001) Vestigial preference functions in neural networks and tungara frogs. Proc Natl Acad Sci USA 98(13):161–166

Richards DG, Wiley RH (1980) Reverberations and amplitude fluctuations in the propagation of sound in a forest: implications for animal communication. Am Nat 115:381–399

Roeder KD, Treat AE (1957) Ultrasonic reception by the tympanic organ of noctuid moths. J Exp Zool 134:127–157

Römer H (1976) Die Informationsverarbeitung tympanaler Rezeptorelemente von Locusta migratoria (Acrididae, Orthoptera). J Comp Physiol 109:101–122

Römer H (1987) Representation of auditory distance within a central neuropil of the bushcricket *Mygalopsis marki*. J Comp Physiol 161:33–42

Römer H (1998) The sensory ecology of acoustic communication in insects. In: Hoy RR, Popper AN, Fay RR (eds.) Comparative hearing: insects. Springer, New York, pp 63–96
Römer H, Spickermann M, Bailey WJ (1998) Sensory basis for sound intensity discrimination in the bushcricket Requena verticalis (Tettigoniidae, Orthoptera). J Comp Physiol A 182:595–607
Schmidt A, Riede K, Römer H (2011) High background noise shapes selective auditory filters in a tropical cricket. J Exp Biol 214:1754–1762
Schmidt A, Römer H (2011) Solutions to the Cocktail party problem in insects: Selective filters, spatial release from masking and gain control in tropical crickets. PLoS ONE 6(12):e28593. doi:10.1371/journal.pone.0028593
Shaw SR (1994) Detection of airborne sound by a cockroach "vibration detector": a possible missing link in insect auditory evolution. J Exp Zool 193:13–47
Smith AR, van Staaden MJ, Carleton KL (2012) An evaluation of the role of sensory drive in the evolution of Lake Malawi cichlid fishes. Int J Evol Biol. doi:10.1155/2012/647420
van Staaden MJ, Römer H (1997) Sexual signaling in bladder grasshoppers: tactical design for maximizing calling range. J Exp Biol 200:2597–2608
van Staaden MJ, Römer H (1998) Evolutionary transition from stretch to hearing in ancient grasshoppers. Nature 394:773–776
van Staaden MJ, Rieser M, Ott SR, Papst MA, Römer H (2003) Serial hearing organs in the atympanate grasshopper Bullacris membracioides (Orthoptera, Pneumoridae). J Comp Neurol 465:579–592
Stavenga DG (2004) Visual acuity of fly photoreceptors in natural conditions: dependence on UV sensitizing pigment and light-controlling pupil. J Exp Biol 207:1703–1713
Wiley RH, Richards DG (1978) Physical constraints on acoustic communication in the atmosphere: implications for the evolution of animal vocalizations. Behav Ecol Sociobiol 3:69–94
Wiley RH, Richards DG (1982) Adaptations for acoustic communication in birds: sound transmission and signal detection. In: Kroodsma DE, Miller EH, Quellet H (eds) Acoustic communication in birds. Academic Press, New York, pp 131–181
Wyttenbach RA, May ML, Hoy RR (1996) Categorical perception of sound frequency by crickets. Science 273:1542–1544
Yack JE, Fullard JH (1993) What is an Insect Ear? Ann Entomol Soc Am 86:677–682
Yack JE, Fullard JH (1990) The mechanoreceptive origin of insect tympanal organs: A comparative study of similar nerves in tympanate and atympanate moths. J Comp Neurol 300:523–534
Yager DD (1990) Sexual dimorphism of auditory function and structure in praying mantises (Mantodea; Dictyoptera). J Zool Lond 221:517–553
Zimmermann U, Rheinlaender J, Robinson D (1989) Cues for male phonotaxis in the duetting bushcricket Leptophyes punctatissima. J Comp Physiol A 164:621–628
Zuk M, Kolluru GR (1998) Exploitation of sexual signals by predators and parasitoids. Quart Rev Biol 73:415–438

Chapter 4
Auditory Parasitoid Flies Exploiting Acoustic Communication of Insects

Berthold Hedwig and Daniel Robert

Abstract Hearing evolved in flies of both Dipteran families *Tachinidae* and *Sarcophagidae*, enabling the parasitic exploitation of singing orthoptera and hemiptera. Guided by acoustic communication signals, these flies identify and localise their singing target, depositing their larvae on or near the host. Larvae then develop as endoparasites, eventually killing the host. In these flies, the mechanosensory auditory organ is located on the prosternum and in most cases is less than one millimetre in size. The frontal facing tympana constitute an extreme example of adaptation in auditory micromechanics. Directional hearing relies on the mechanical coupling between the hemilateral tympana, a purely mechanical process that exploits minute interaural time differences in tympanic vibrations and enhances bilateral oscillation differences to generate a highly directional sensitivity. In tachinid fly species, the frequency response of the ears is adapted to the host communication signals. The auditory organs contain up to 250 scolopidial afferents, which are directly driven by tympanic membrane vibrations. The signals from the auditory afferents are forwarded to auditory neuropils in the three thoracic neuromeres. Further processing of intensity and directional information and also of temporal patterns involves local and also ascending auditory interneurons, which project up to the brain for final sensory-motor integration.

B. Hedwig (✉)
Department of Zoology, Cambridge, UK
e-mail: bh202@cam.ac.uk

D. Robert
School of Biological Sciences, Bristol, UK
e-mail: d.robert@bristol.ac.uk

B. Hedwig (ed.), *Insect Hearing and Acoustic Communication*,
Animal Signals and Communication 1, DOI: 10.1007/978-3-642-40462-7_4,

4.1 Introduction

Acoustic signalling rarely constitutes a private communication channel. Sound tends to radiate in all directions, propagating towards hardly predictable destinations and, sometimes, unintended receivers as predators and parasitoids. Some flies take advantage of acoustic signalling in insects to feed their larval brood. Hearing evolved among the Dipteran parasitic family Tachinidae and the flesh flies family Sarcophagidae, which are to date the only Dipteran Brachycera known to use acoustic cues to localise their host (Leonide 1969; Cade 1975; Soper et al. 1976; Fowler 1987, Barraclough and Allen 1996, Robert et al. 1998). Infestation relies on the fly planidia larvae to actively burrow their way into the host. In some species, such as *Ormia depleta*, the larvae are deposited at some distance from the host (Wineriter and Walker 1990), around the entrance of a mole cricket burrow from which the song is broadcasted. The larvae then make contact with the host by frictional encounter, or by crawling towards it. The larvae develop inside the host, a process that results in the host's death and the egress of 3–9 larvae.

There seems to be a definite advantage to find a host free of previous parasitic load; field observations of tachinids suggest that being first may be a matter of seconds. Fast and efficient host finding using acoustic cues appear to constitute a key element in the reproductive biology of Dipteran parasitoids. In populations with high infestation rates, parasitoid flies constitute significant natural selection, having a direct effect on the intraspecific communication of their hosts (Cade 1975, Walker 1993, Adamo et al. 1995a, Zuk et al. 2006).

The auditory organs of flies exhibit microscale tympanic membranes that are located at the prosternum, on the ventral prothorax at the base of the neck (Lakes-Harlan and Heller 1992, Robert et al. 1992). Insects of the order Diptera tend to be small; endoparasitic Diptera are constrained to be even smaller than their hosts, a limitation that has direct consequences for their sense of hearing. In effect, a small distance between the ears severely limits the use of interaural sound amplitude differences as directional cues. Highly acute auditory orientation, however, is possible due to an auditory system endowed with mechanically coupled tympanal membranes (Robert et al. 1994; Robert 2001; Mason et al. 2001) and appropriate neural processing at the level of auditory afferents and interneurons (Oshinsky and Hoy 2002; Stumpner et al. 2007).

Hearing research in insects has shown the diversity of miniature solutions to the problems of sound reception (Robert and Hoy 2007) and processing, (Hennig et al. 2004; Hedwig and Pollack 2008), illustrating the often unexpected routes of sensory evolution (Fullard and Yack 1993; Montealegre-Z et al. 2012). In this chapter, we review the co-evolutionary relationship between parasitoid flies and their insect hosts, focussing on the flies phonotactic behaviour (Müller and Robert 2001; Ramsauer and Robert 2000), their micro-scale hearing system (Miles et al. 1995, Robert et al. 1996) and the neural organisation of their auditory pathway (Oshinsky and Hoy 2002; Stumpner and Lakes-Harlan 1996; Stumpner et al. 2007).

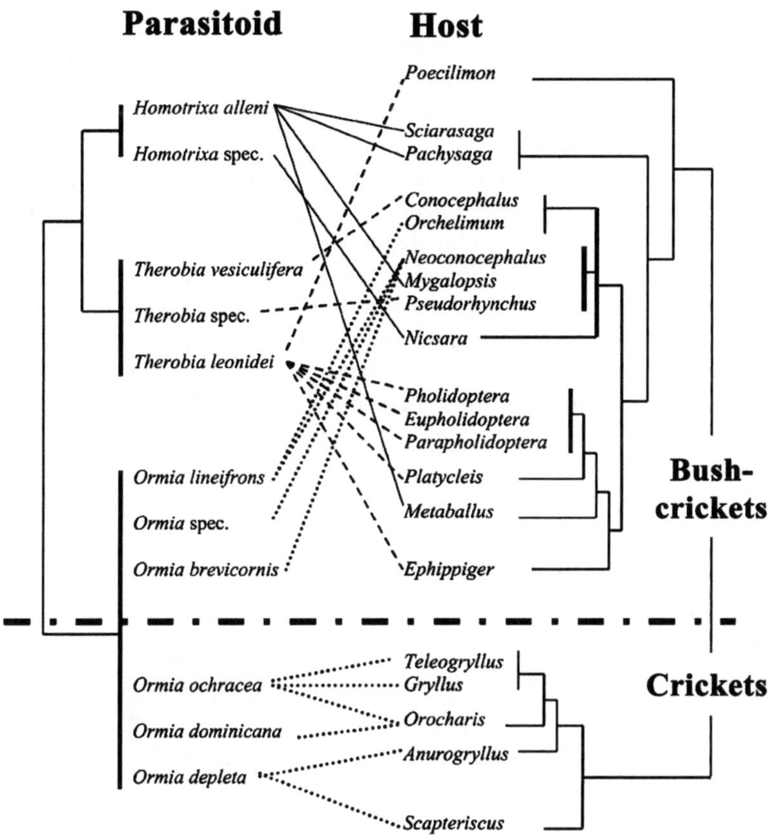

Fig. 4.1 A phylogenetic tree of parasitic Tachinid flies and their bushcricket and cricket host species for 11 Ormiini flies. Only Ormia species in North America target crickets as hosts; modified from Lehmann (2003, Fig. 4.2)–with kind permission of the author

4.2 Parasitic Hearing Flies and Hosts

Among the ~8000 species of Tachinid flies there are 67 known species in the group of parasitoid Ormiine that use acoustic cues to localise their host. Ormiini are found in tropical and subtropical regions where at least 11 species prey on various orthopteran insects. *Therobia leonide*, the only Ormiine fly in southern Europe (Léonide 1969; Lakes-Harlan and Heller 1992; Lehmann et al. 2001) and the Australian species *Homotrixa allen*, (Fig. 4.2a) (Allen et al. 1999) target bushcrickets. Further genera, such as *Phasioormia*, *Euphasiormia*, *Aulacephala*, *Mediosetiger* have been examined from museum collections and show to bear auditory organs (Huber and Robert, unpublished), however, host species are not yet known. Field collection using acoustic trapping (Wineriter and Walker 1990) has not yet been carried out across the bandwidth of Orthopteran songs. Such field prospection may uncover a greater diversity of acoustic parasitoids, and live material to examine.

In North America (Florida) gravid *Ormia lineifrons* heavily parasitises *Neoconocephalus triops* (Burk 1982); *Ormia depleta*, a species originating from South America, is attracted to the song of mole crickets of the genus *Scapteriscus* (Fowler and Kochalka 1985) and *Ormia ochracea* (Fig. 4.2b) targets singing field crickets; exhibiting regional differences in their host preferences. *O. ochracea* targets *Gryllus rubens* in Florida and in western USA it parasitises the variable field cricket *Gryllus lineaticeps* (Cade 1975). *O. ochracea* has been introduced to the Hawai'ian archipelago where it targets the Polynesian field cricket *Teleogryllus oceanicus,* another introduced species (Zuk et al. 1993). Overall, the flies seem to be opportunistic hunters with a wide range of potential hosts as auditory pattern recognition in flies is not as specific as intraspecific communication shaped by sexual selection. Lehmann (2003) speculates that bushcrickets were the hosts in the ancestral form of parasitism and that the shift to Gryllids is the derived state as these are only targeted by the North American genus *Ormia*.

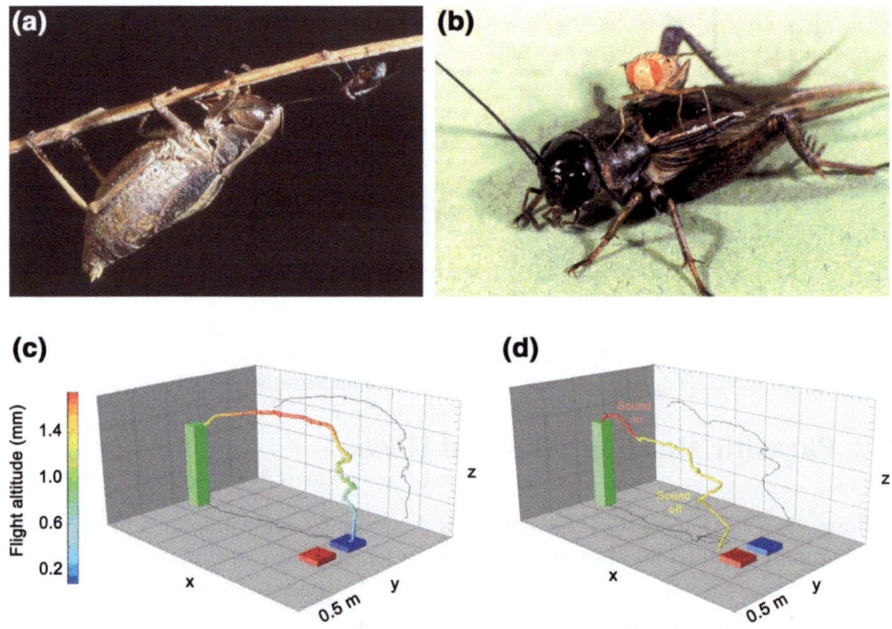

Fig. 4.2 **a** The Australian Tachinid fly *Homotrixa alleni* with its bushcricket host *Sciarasaga quadrata* in the wild. **b** The North American *Ormia ochracea* depositing larvae on her host photographed in the laboratory. **c, d** Phonotactic behaviour of free-flying *O. ochracea* towards a loudspeaker broadcasting a cricket song. Green block is the starting platform; red and blue are landing platforms each hosting one loudspeaker. **c** Phonotaxis to blue loudspeaker, red speaker is off. **d** Same fly attracted to red loudspeaker presenting cricket song (red trajectory) and its flight path after the song was stopped (yellow trajectory). Photo in **a** courtesy by Geoff Allen, University of Tasmania/Hobart and photo in **b** courtesy by Marie Read, Cornell University; **c** and **d** modified from Müller and Robert (2001)

Parasitoid Tachinid flies are strictly crepuscular and nocturnal with phonotaxis occurring in gravid females only. These land next to the singing host and walk towards it, dropping planidia, the mobile first instar larvae directly on the host and/or in its immediate vicinity (Cade 1975; Adamo et al. 1995b; Allen et al. 1999). In general, only singing male hosts are infested but if females respond acoustically to calling males, they can also suffer parasitism. The extension of parasitism to non-singing females arises from the flies' larviposition strategy. In *O. ochracea*, three larvae were directly deposited on the host, whilst more planidia were laid around the host, ambushing female crickets approaching the calling male. Upon contacting the host, the larvae climb on it and crawl to find soft intersegmental cuticle, through which they bore an entry point. Once inside the host, they feed on hemolymph and muscle, but keep the host alive and alert until the last stages of parasitism. Infested crickets die after 9–12 days (Walker and Wineriter 1991; Adamo et al. 1995a), upon the egress of the maggots.

Acoustically guided parasitism is also found among the *Sarcophagidae* and represents a case of a convergent evolution within the Diptera (Lakes-Harlan et al. 1999; Robert et al. 1999). In North America the sarcophagid fly *Emblemasoma sp.* and *Emblemasoma auditrix* mainly parasitises the calling males of cicada, e.g. *Okanagana rimosa* that sing during daytime (Soper et al. 1976; Farris et al. 2008). Flies first land next to the singing host and a sequence of highly specialised behaviour follows that is distinct from that generally observed in Tachinids. After visual orientation towards the host's abdomen, the fly squeezes under the cicada's wings, and proceeds to cut through the timbal organ, damaging it using their genital plate. The fly then deposits a larva within the sound-producing organ before flying away (Schniederkötter and Lakes-Harlan 2004). With a damaged sound-producing organ, male singing is less prevalent and the infected cicada deemed less likely to attract additional flies. This sophisticated parasitic strategy prevents superparasitism and therefore larval competition. Exposure to parasitoids may also be moderated by the time of day at which singing takes place. It was suggested that the different dial singing activity of three sympatric cicada species may reflect different avoidance strategies to reduce the risk of parasitism by day-active *Emblemasoma sp.* (Farris et al. 2008).

4.2.1 Host Selection and Communication Signals

Sexual selection is considered the primary force that shaped intraspecific acoustic communication signals in insects. In contrast, in parasitoid flies, auditory signal preferences with respect to sound frequency and temporal patterns have to be due to natural selection. The carrier frequency of the communication signal alone does not determine host selection (Ramsauer and Robert 2000). The processing of temporal patterns appears to be less sharply tuned as in the host species, but demonstrates some adaptation to the signal structure of the host.

In central North America (Michigan) the Sarcophagid *E. auditrix* specifically parasitises the cicada *O. rimosa*. For phonotactic approaches, *E. auditrix* favours a pulse repetition rate of 80 Hz corresponding to the chirp rate of the cicada's calling song (Köhler and Lakes-Harlan 2001; Schniederkötter and

Lakes-Harlan 2004). In more southern regions (Mississippi) *Emblemasoma sp.* is more attracted to *Tibicen pruinosa*, one of three sympatric cicada species (along with *T. chloromera*, and *Neocicada hieroglyphica*) with a pulse rate of about 1.5 Hz (Farris et al. 2008).

In the Tachinid flies *Homotrixa alleni* and *Therobia lenonidei*, host selection depends on the call structure of the host (Allen 2000; Lehmann and Heller 1998). In populations of the bushcricket *Sciarasaga quadrata* males with shorter chirps (55.4 ms versus. 64.8 ms) and higher chirp rates (112 Chirps/min versus 103 Chirps/min) were lost from the calling population due to parasitisation. In Greece, the closely related species *Poecilimon mariannae* and *Poecilimon veluchianus* generate 5–11 or just a single pulse per chirp, respectively. In experimentally mixed populations of both species *T. leonidei* infested the polysyllabic species about 3.6 times more often. However, such infestation probabilities do not necessarily reveal a preference for a specific song pattern, but rather reflect that songs with a higher pulse rate allow for a higher sampling of the host's location during the fly's approach (Lehmann und Heller 1998).

Across North America the acoustic ecology of parasitoid Ormiini may be more complex as gravid females of *O. ochracea* are attracted to different cricket songs in different regions of the US. This situation was tested by Gray et al. (2007) who exposed geographically separated *Ormia* populations in Florida, Texas, California, and Hawaii to the song models of 4 cricket species (*Gryllus rubens*, *G. texensis*, *G. lineaticeps* and *Teleogryllus oceanicus*) representing the most common host species in the study areas. These species have a carrier frequency in the range of 4–5 kHz in common but produce songs with highly different temporal patterns (Fig. 4.3). *O. ochracea* females exhibited clear local preferences and in Florida were most strongly attracted to *G. rubens* song (Walker 1993), in Texas to the song of *G. texensis*, in California to *G. lineaticeps* and to *T. oceanicus* in Hawaii. The flies' geographically indicate specific responses a host-specific adaptation of auditory processing, which may have been shaped during co-evolution of host and parasite and possible competition between sympatric fly species.

4.2.2 Impact of Parasitic Flies on Host Communication Systems

Cade (1975) suggested that predation by parasitic flies will act as a natural selection pressure on male reproductive behaviour. The impact of acoustic parasitoid flies on cricket populations could be closely studied on the Hawaiian archipelago where flies (*O. ochracea*) and the Polynesian field cricket (*T. oceanicus*) were recently introduced to three of the islands (Zuk et al. 2006). On the island of Kauai about 30 % of singing males were found to be infested by the flies. Notably, in the late 1990s, within about 5 years and less than 20 generations a "flat-wing" mutation spread among the cricket population. It left 90 % of the males mute, due to a reduced stridulatory apparatus with a strongly shortened and

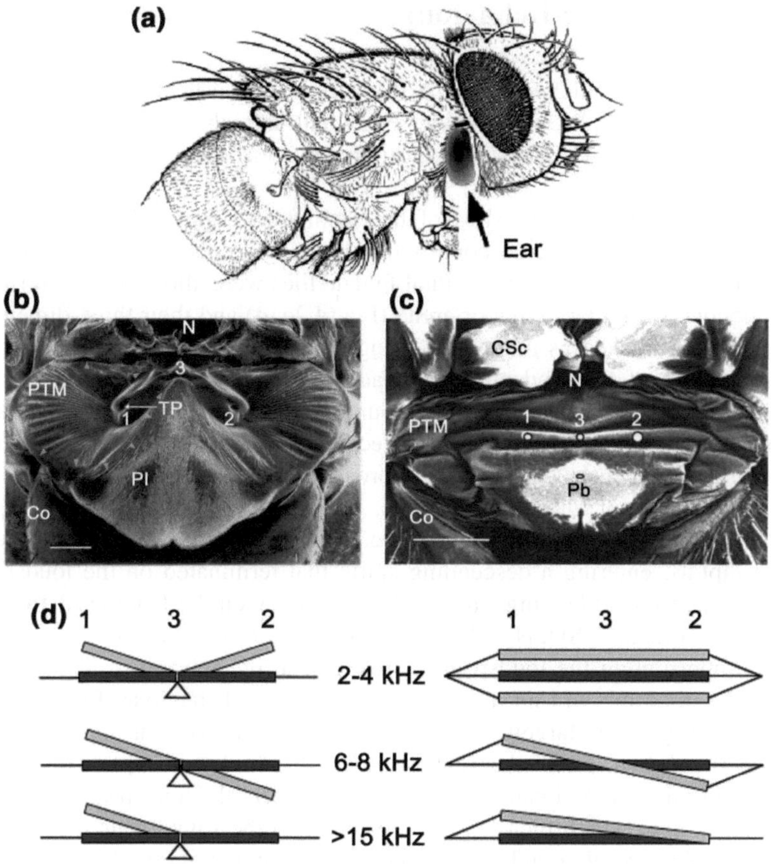

Fig. 4.3 **a** Head and thorax of a generalised Tachinid fly indicating the position of the ear at the prosternum. **b** Scanning electron micrograph of the tympanal ear of *O. ochracea* and **c** the ear of the Sarcophagid fly *Emblemasoma spp* with PTM: prosternal tympanal membrane. TP: tympanal pit of the presternum; PI: prosternal inflation; N: neck insertion; Co: prothoracic coxa and CSc: cervical sclerite; Pb: probasisternum. 1, 2 indicate attachment site of auditory afferents and 1–3 the position of laser vibrometric measurements. **d** Deflection modes of the tympanal system in *Ormia* (left) and *Emblemasoma* (right) measured at positions 1–3. The bars represent the intertympanal bridge in Ormia and the horizontal tympanal fold in Emblemasoma. Tympanic membranes lead to frequency-dependent oscillations of the attachment sites of the sensory cells, with a maximum response at 6–8 kHz. (Scale bar B: 200 μm and C: 500 μm. **a** from Yager (1999) with permission of John Wiley and Sons; **b** from Robert et al. (1994); **c, d** from Robert et al. (1999)

relocated file, precluding sound production. Flat-winged males cannot call and become satellite males, in the vicinity of the remaining active singers. Such mute males run a lower risk of direct fly parasitism; their rate of infestation was less than 1 %. Thus, the predation pressure of the flies established and stabilised the presence of a mutation in the cricket population that in normal populations would be detrimental (Zuk et al. 2006; Tinghitella et al. 2009).

4.3 Host Targeting Behaviour

In their search for a suitable host, the flies need to acoustically identify their target, and precisely localise its position. For small animals with a miniature brain, such a sensory task and behaviour constitutes a demanding challenge of auditory processing.

The behavioural characteristics of the phonotactic approach of *O. ochracea* were investigated using stereoscopic video tracking cameras (Müller and Robert 2001, 2002; Fry et al. 2000). Individual female flies were allowed to perform free-flight phonotaxis in a large indoor arena (Fig. 4.2c, d) and their three-dimensional flight path was tracked as they were flying in complete darkness from a starting platform to one of the loudspeakers broadcasting a cricket song from the floor. While the flies approached the sound source the delivery of the acoustic stimulus could be manipulated. The flies' trajectory was not following a straight line between start and finish. Instead they approached the sound source first maintaining a rather constant altitude (Fig. 4.2c, blue loudspeaker), following a gently meandering trajectory. At some point directly above the sound source, the flies lost altitude rapidly, entering a descending spiral that terminated on the loudspeaker. When repeated for 80 landings, the landing accuracy on the landing platform was 8.2 ± 0.6 cm (mean $\pm$ SD; $N = 10$ flies; Müller and Robert 2001).

When interrupting the loudspeaker broadcast at predetermined distances from the sound source this did not disrupt the phonotactic behaviour (Fig. 4.2d). The phonotactic flight was largely conserved exhibiting the typical approach pattern and final spiral dive. This behaviour is remarkable in that it implies that the flies acquired sufficient directional information as to where the loudspeaker was *before* song interruption, which then led the flies close to the sound source in darkness, even in the absence of sound. Furthermore, this result suggests the presence of a form of directional memory or at least sequential processing of acoustic signals, and its retrieval for non-idiothetic navigation.

After landing next to a host *O. ochracea* also show walking phonotaxis (Cade 1975; Walker 1993) whereby orientation and motion take place only while sound is present. The accuracy of walking phonotaxis has been tested on a treadmill and shown to be remarkably accurate, as flies localise sound sources to within 2° azimuth; revealing their acute sense of directional hearing (Mason et al. 2001).

These results highlight potential limitations of the cricket's song interruption strategy for the avoidance of acoustic parasitism. From the cricket's perspective the question arises as to what sensory capacity may be involved to warn of an approaching fly. Using the cercal system, field crickets are deemed to be capable of detecting the sounds of approaching flying or walking predators. Sensitive to the particle velocity component of a sound wave, as well as bulk motion of air (Gnatzy and Heusslein 1986; Dangles et al. 2006) the cercal system of the cricket functions as an alarm, which can trigger the interruption of singing (Dambach and Rausche 1985) and may avoid parasitism (Müller and Robert 2002).

4.4 The Ears of Parasitoid Flies

The auditory apparatus of all parasitoid flies are located on the ventral prothorax, they are characterised by a modified inflated prosternum (Fig. 4.3a; Shewell, 1987) and represent an example of convergent evolution within the Diptera (Lakes-Harlan et al. 1999; Robert et al. 1999).

4.4.1 Auditory Morphology and Comparison Between Species

In the Tachinid flies *T. leonidei* and *O. ochracea* both sexes have a balloon-like inflated prosternum filled with an air sac, however, the ears are sexually dimorphic, with the female ears larger than the male's. The ears of Ormiine flies consist of a bladder-like prosternum between the fore coxae and the head capsule (Fig. 4.3b; Robert et al. 1992, 1994). In at least two species, namely *O. ochracea* and *T. leonidei*, best auditory sensitivity is similar—but not identical—to the peak of the frequency spectrum of the host song (Lakes-Harlan and Heller 1992; Robert et al. 1992). The sensory organs proper each are enveloped in tracheal tissue and connect the sternal apophysis with the cervical membrane. The sensory organ consists of about 200 scolopidial sensory cells. The axons of the receptor cells run within a short nerve and enter the thoracic-abdominal ganglion where they form arborisations in all three thoracic neuromeres.

Hearing sarcophagids of the genus *Emblemasoma* (Robert et al. 1999, Farris et al. 2008) are parasitoids of cicadas. The bilateral tympanal membranes of the auditory organ present a series of folds spanning from one side of the prosternum to the other (Fig. 4.3c). In the sarcophagid *E. auditrix*, the bilateral tympanal membranes are median separated by sclerotised cuticular structures. The tympanal membranes extend laterally towards the coxae. Like in Ormiine flies, a single air sac fills the space behind the tympanal membranes. On each side one scolopidial organ is attached to the presternum of one tympanal membrane and the corresponding prosternal apodeme. The sense organ is located close to the apodeme and contains about 30 scolopidia, each with a sensory cell, a scolopale cell and accessory attachments cells. From the auditory organ, the auditory nerve runs caudally and the axons of the auditory afferents project into thoracico-abdominal ganglion complex. The prosternal tympanal membranes of *E. auditrix* exhibit mechanical oscillations when exposed to sound stimuli in the range of 3–30 kHz, a possible indication of its auditory sensitivity in that frequency range.

In the non-hearing atympanate Sarcophagids (e.g. *Sarcophaga bullata, Phormia regina*) a prosternal chordotonal organ in the ventral prothorax is deemed to represent the evolutionary precursor of the hearing organs (Edgecomb et al. 1995; Lakes-Harlan et al. 1999, 2007; Stölting et al. 2007). This chordotonal organ contains about 35–55 mechanosensitive afferents, which in *S. bullata* respond in a phasic-tonic way to substrate vibrations in the range of 0.2–4 kHz (Stölting et al.

2007) transmitted via the legs. It may be regarded as a pre-adaptive structure from which hearing organs independently evolved in Sarcophagids and Tachinids. In such scenario, modifications of the peripheral cuticular structures of the prosternum resulted in the sensitivity of the prosternal mechanosensory organ to shift from leg-vibrations to vibrations of the prosternal membranes. As they would evolve to be thinner, these membranes could become mechanically sensitive to sound waves, as tympanal membranes are.

Interestingly, the prosternal membranes of Tsetse files are as thin as tympanal membranes but do show some mechanical sensitivity to sound (Tuck et al. 2009). As such the prosternal region of *Glossina* might have been suggested to act as a hearing organ, as it shows some of the characteristics required. Altogether, these results may be used to hypothesise that, following this plausible evolutionary scenario, some parasitoid (or parasitic) flies employ the plesiomorphic prosternal chordotonal organ to locate their hosts using substrate vibrations.

4.4.2 Biomechanics of the Dipteran Tympanal Ears

The tympanal ears of parasitoid flies are in the majority of species less than 1 mm wide and from physical acoustics meet some limitations for processing incoming sound waves. The possible functional issues related to reduce mechanical sensitivity due to the small surface area finite thinness of tympanal membranes, mechanical stability of the entire apparatus if situated on or near moving parts, and as expected, problems with the directional detection of incident sounds. The problem of directional hearing using microscale hearing organs has been given some attention, in particular in the fly *O. ochracea* (Robert and Göpfert 2002; Robert 2005). Because the ears are set so close together on the prothorax of the flies, the acoustic information they can derive is vanishingly small, but not negligible. In Ormia, conforming to conventional acoustic cues used for directional hearing, the maximal amplitude and time differences experienced by each tympanal membrane amount to a fraction of 1 dB in amplitude and 1.45–2 μs (Robert et al. 1996), respectively—this is the best-case scenario. The time differences become even smaller when the sound source is situated near the front of the animal, yielding interaural time differences in the submicrosecond range (Robert 2001; Mason et al. 2001).

Directional hearing in tachinid flies is made possible by the capture of acoustic energy with tympanal membranes that are mechanically coupled (Robert et al. 1996; Robert 2005, 2008). Mechanical coupling is responsible for the generation of temporal and amplitude mechanical cues large enough for the directional processing of information by mechanosensory neurones in each of the auditory organs (Robert et al. 1996; Mason et al. 2001). The anatomical structure responsible for intertympanal coupling is the presternum, a sclerite that is present in most brachyceran Diptera but has developed only a particular morphology and functionality in Tachinid flies. The presternum constitutes a flexible mechanical link between the two tympanal membranes (Fig. 4.3b, c). It is also the point of attachment of the mechanosensory organs, thus presternal vibrations will determine the input to the sensory organ (Robert et al. 1992, 1996). Biomechanical studies have shown that the hemilateral branches of the presternal, the cuticle linking the tympanal pits (TP, Fig. 4.3b, c) act like a flexible see-saw, rocking back and forth in the sound field (Robert et al. 1996). At frequencies at and slightly above that of the song of the cricket host, the presternum acts as a somewhat floppy cuticular beam, resulting in a rocking motion whereby both ends of the beam do not vibrate at the same time and magnitude: The side nearer the sound source oscillates with a slight phase lead, and larger amplitude (Fig. 4.3d).

In Sarcophagid flies, intertympanal coupling has been shown to also be the basis of directional sensitivity (Robert et al. 1999). Several anatomical differences have been identified in the organisation of the tympanal membranes (Fig. 4.3). Although structurally different, the physical linkage between the two tympanal membranes was shown to generate the coupling necessary for directional hearing. The main difference is that a transversal fold on the tympanal membranes produces the stiffness necessary to the mechanical coupling, notably obviating the need for a central fulcrum point (Robert et al. 1999).

The functional properties of fly ears have prompted the development of miniature microphones, which use the physical principles discovered in flies to generate unprecedented directional sensitivity (Miles and Hoy 2006; Miles et al. 2009).

4.5 Organisation of the Auditory Pathway and Auditory Processing

The overall organisation of the central auditory pathway in flies resembles that in other insects (Stumpner and Helversen 2001; Hedwig and Pollack 2008). In all auditory flies hearing is based on scolopidial sensilla, the axons of the mechanosensitive neurons, which constitute the auditory afferents, leave the hearing organ to enter the thoracic-abdominal through a prothoracic nerve (Fig. 4.4a–c) and project to all three thoracic auditory neuropils; ascending interneurons forward auditory activity towards the brain.

4.5.1 Auditory Afferents

In Tachinid flies the total number of auditory afferents is 90–100 for *O. ochracea* (Oshinsky and Hoy 2002), about 250 for *T. lenonidei* (Lakes-Harlan et al. 2007), and in *H. alleni* there are about 200 with two groups of scolopidia considering their position and diameter (Fig. 4.5b; Stumpner et al. 2007). The number of sensory neurons is in the range of 30 for the Sarcophagid *E. auditrix* (Lakes-Harlan et al. 1999).

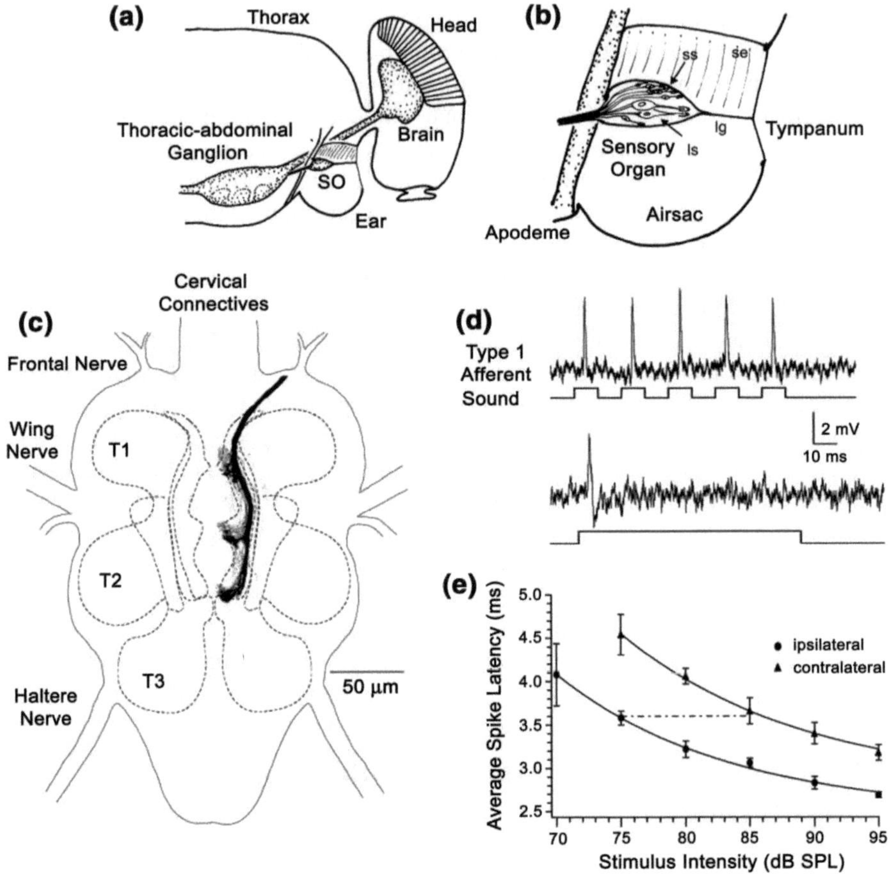

Fig. 4.4 **a** Schematic anatomy of the tympanal organ of the tachinid *H. alleni* with the location of the prosternal ear in relation to the brain and the thoracic-abdominal ganglion. SO: auditory sensory organ. **b** Details of the ear showing large (ls) and small (ss) types of scolopidia; se: septum; lg ligament. **c** The central projections of the auditory nerve in *O. ochracea* end in the ventral regions of the three thoracic neuromeres. **d** Response of a type 1 auditory afferent to synthetic cricket song with 50 pulses per second at 5 kHz, 85 dB SPL (top) and to a 100 ms pulse at 7 kHz/80 dB SPL (bottom). **e** Direction-dependent response of type 1 afferents. A latency difference occurs for identical sound intensities presented from either side of the animal. The dashed line indicates that the shift in latency due to ipsilateral versus contralateral stimulation corresponds to a 10 dB difference in stimulus intensity. **a**, **b** from Stumpner et al. (2007, Fig 4.1)-with kind permission of Springer Science+Business Media, **c**–**e** from Oshinsky and Hoy (2002) with permission of Journal of Neuroscience

The axons of the auditory afferents project from the auditory organ into the fused thoracic-abdominal ganglion complex forming a well-defined track (Figs. 4.4c, 4.5a). Axons arborise ipsilateral in all three thoracic neuromeres in the medial Ventral Association Center (mVAC). In insects in general, this is the neuropil area which receives a variety of mechanosensory inputs and which is also the projection target for auditory afferents (Stumpner and Helversen 2001; Hedwig and Pollack

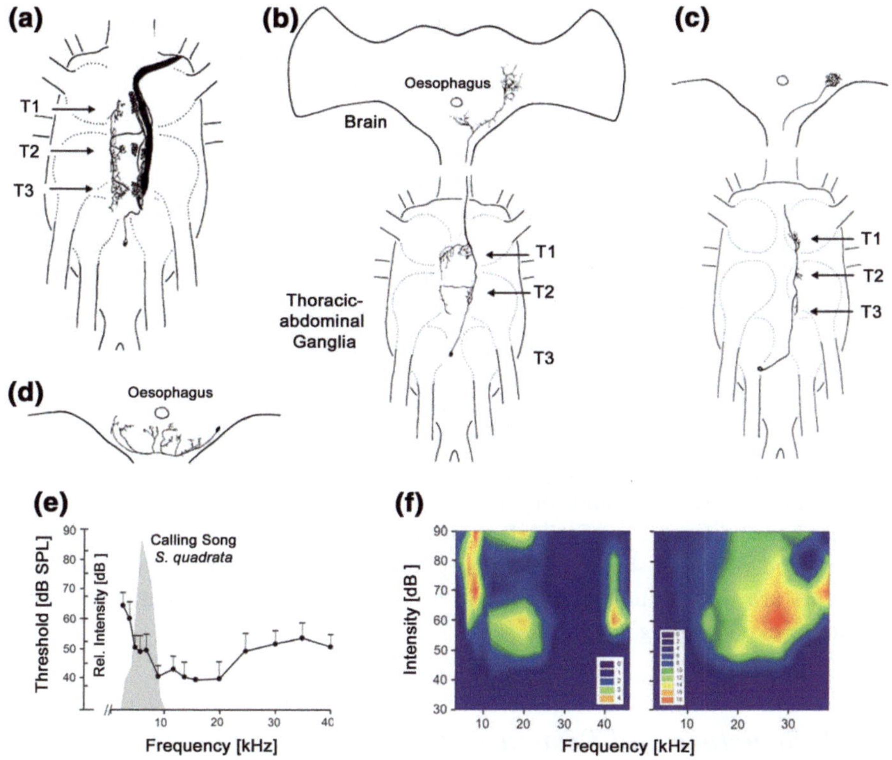

Fig. 4.5 Auditory pathway in *H. alleni*. **a** Projection pattern of auditory afferents in the thoracic neuromeres T1–T3 *right* and structure of a local bilateral auditory interneuron *left* stained via dye coupling from the sensory fibres. **b**, **c** Morphology of two ascending auditory interneurons projecting to the brain; arrows point to dendritic arborisations in the thoracic neuromeres; axonal structures are in the deutocerebrum. **d** A local bilateral auditory interneuron in the suboesophageal ganglion. **e** Frequency tuning of the ascending auditory pathway as recorded from the neck connectives plotted against the power spectrum of the host *S. quadrata left*. **f** Intensity-frequency plots, demonstrating the response properties of a low frequency interneuron and a broadly tuned interneuron. Red colours indicate high spike activity; note different scaling for both neurons. **a–f** modified from Stumpner et al. (2007, Figs. 4.3, Fig. 4.4b, Fig.7)-with kind permission of Springer Science+Business Media

2008). No afferents have been observed to project directly to the brain or into the abdominal neuromeres (Oshinsky and Hoy 2002; Lakes-Harlan et al. 1999).

The response properties of auditory afferents in *Ormia* appear to be adapted to processing of directional and patterned acoustic signals (Oshinsky and Hoy 2002). Although axon diameters in *Ormia* are less than 2 μm single cell recordings identified 4 types of afferents, two of which exhibit highly phasic response properties, generating only 1 or very few spikes in response to 5 kHz (80 dB SPL) sound pulses (Fig. 4.4d). Spikes as recorded in the thoracic-abdominal ganglion occur with a mean latency of 2.8 ms and moreover with an extremely low jitter of only 76 μs allowing a precise time coding of stimulus onsets, as required for acute directional processing. Upon stimulation from 90 deg ipsilateral or contralateral Type 1 afferents exhibit an intensity-dependent difference in response latency of about 0.5–1 ms (Fig. 4.4e). The shift in the latency response curves for ipsilateral and contralateral stimulation equals to a difference in sound intensity of about 10 dB, matching a corresponding difference of tympanal membrane vibrations (Robert et al. 1996).

At the population level threshold intensities spread over a wide range from 50 to 93 dB SPL with a nearly linear increase in the number of activated afferents between 68 and 88 dB SPL. Bilateral differences in the amplitude of tympanic membrane vibrations (Robert et al.1996) will therefore recruit different number of auditory afferents on both sides of the animal, supporting directional coding by means of latency differences and intensity-dependent afferent recruitment. This effect was previously reported for Orthopteran species (Römer et al. 1998), whereby latency differences may further be processed and enhanced by local auditory interneurons in the central nervous system. Coupled to the low jitter of the afferent responses is an extremely long refractory period of about 4.3 ms (Oshinsky and Hoy 2002). Thus, these auditory afferents are well suited to respond in a timely manner to the patterned structure of crickets' calling songs.

4.5.2 Structure and Response Properties of Auditory Interneurons

There is gradually increasing information on the neural basis of central auditory processing. Morphology and response properties of auditory interneurons have been analysed in the Tachinid flies *T. lenonidei* and *H. alleni* (Stumpner and Lakes-Harlan 1996; Stumpner et al. 2007). In the thoracic ganglia, dendritic arborisations of local interneurons overlap and intermingle within the axonal projections of the afferents in all three neuromeres. A bilateral arborisation pattern of some local interneurons may contribute to directional processing. At least some of the local interneurons appear to be electrically coupled to the auditory afferents, as dye—coupling during backfills of the auditory nerve also revealed the structure of the local interneurons (Fig. 4.5a–d).

Some interneurons of the ascending auditory pathway have been identified; these have a cell body in the pro, meta, or abdominal neuromere. Their dendrites overlap to different degrees with one or both sides of the auditory neuropils in the mVAC of the thoracic neuromeres. Axons of the interneurons project—in most cases contralaterally to the cell body—towards the suboesophageal ganglion and the brain. Axonal arborisations spread out in the suboesophageal ganglion (*Homotrixa*) and typically in the lateral deutocerebrum. Whereas some interneurons have wide-field projections others form a dense glomerulus-like axonal projection pattern in the brain (Fig. 4.5b, c). Overall interneurons cover a similar neuropil area in the deutocerebrum, pointing towards a specific dedicated area for auditory processing within the brain. At least one bilaterally projecting auditory neuron has been characterised in the suboesophageal ganglion; no local auditory brain neurons have been identified to date. The structure of fly auditory interneurons is highly similar in *Homotrixa* and *Therobia*, pointing towards homologous neurons involved in auditory processing in both species (Stumpner et al. 2007).

In *T. leonidei,* the ascending auditory pathway is broadly tuned to 16–40 kHz and matches the power spectrum of the bushcricket host's calling song (Stumpner and Lakes—Harlan 1996). Female thresholds were about 45 dB SPL whereas male flies were 10–20 dB less sensitive. As all interneurons in *T. leonidei* had similar tuning curves, there is no evidence for frequency discrimination.

In *H. alleni,* hearing sensitivity is similar in female and male flies. Ascending interneurons revealed an overall broad tuning (Fig. 4.5f) with a 50 dB threshold occurring between 5 and at least 40 kHz, and the lowest threshold of 40 dB SPL in the range of 10–20 kHz (Stumpner et al. 2007). Tuning of the ascending interneurons demonstrates no specific adaptation to the carrier frequency of the main host *S. quadrata*, which is at 5 kHz (Römer and Bailey 1998). However, at some interneurons give a strong response at 5 kHz when stimulated at high sound intensities (80–90 dB SPL). Interneurons differ in their frequency tuning and intensity response functions (Fig. 4.5f) indicating the possibility for frequency discrimination as a basis for frequency dependent host discrimination as *H. alleni* can also parasitize bushcricket species like *Mygalopsis pauperculus* and *Pachysaga croceopteryx* (Barraclough and Allen 1996).

In *T. leonidei* temporal processing in ascending interneurons revealed a variety of tonic, phasic and phasic-tonic response properties (Stumpner and Lakes-Harlan 1996). Shortest response latencies were only 5–10 ms and phasic spike rates as high as 500–900 AP/s. Temporal responses depended on frequency and intensity of the stimuli. As *Therobia* parasitises a range of bushcrickets with pulse rates ranging from 0.3 to 10 Hz local fly populations may be adapted to different song patterns of their hosts.

4.5.3 Sensory-Motor Integration

The neural networks for host identification and the descending networks controlling the flies' directional phonotactic approach are still unknown. It seems, however, that the neural mechanisms underlying host selection allow for an evolutionary rapid adaptation as different populations of the same species of flies may select hosts with very different acoustic signals. Earlier stages in auditory processing, including the mechanical response of the coupled ears, seem to be conducive to such flexibility. In effect, the tympanal response to sound can be directional across a large range of frequencies and easily respond to short sound transients. This part of the mechanical response therefore is not a key constraint for the specificity of host finding. Rather, higher level signal processing dealing with temporal species-specific signatures, seem to play a role.

4.6 Conclusion

Parasitoid flies show remarkable specific adaptations in their auditory host-seeking sensory equipment yet also employ opportunistic adaptation to diversify their access to their orthopteran hosts. Further studies in the physiological compatibility of parasitoid flies to their hosts and close relatives could guide further research on the process of co-evolution between singing insects and their acoustic parasitoids. To date in the context of a parasitoid life history the use of hearing has not been reported for insects other than Sarcophagid and Tachinid dipterans. Considering a broader parasitic life history, the haematophagous culicid dipteran *Uranoteania* and *Corethrella* are suggested to use their antennal hearing organ to detect and locate its frog hosts (Borkent and Belton 2006; Bernal et al. 2006).

References

Adamo SA, Robert D, Hoy RR (1995a) Effects of a tachinid parasitoid, O. ochracea, on the behaviour and reproduction of its male and female field cricket hosts (Gryllus spp). J Insect Physiol 41:269–277

Adamo SA, Robert D, Perez J, Hoy RR (1995b) The response of an insect parasitoid, *Ormia ochracea* (Tachinidae), to the uncertainty of larval success during infestation. Behav Ecol Sociobiol 36:111–118

Allen GR (2000) Call structure variability and field survival among bushcrickets exposed to phonotactic parasitoids. Ethology 106:409–423

Allen GR, Kamien D, Berry O, Byrne P, Hunt J (1999) Larviposition, host cues, and planidial behavior in the sound-locating parasitoid fly Homotrixa alleni (Diptera: Tachinidae). J Insect Behav 12:67–79

Barraclough DA, Allen GR (1996) Two new species of Homotrixa Villeneuve (Diptera: Tachinidae: Ormiini) from Southwestern Australia, with data on biology and ecology. Aus J Entomol 35:135–145

Bernal XA, Rand S, Ryan MJ (2006) Acoustic preferences and localization performance of blood-sucking flies (*Corethrella*, Coquillett) to túngara frog calls. Behav Ecol 17:709–715

Borkent A, Belton P (2006) Attraction of female *Uranotaenia lowii* (Diptera: Culicidae) to frog calls in Costa Rica. Can. Entomol. 138:91–94

Burk T (1982) Evolutionary significance of predation on sexually signalling males. Fla Entomol 65:90–104

Cade W (1975) Acoustically orienting parasitoids: Fly phonotaxis to cricket song. Science 190:1312–1313

Dambach M, Rausche G (1985) Low-frequency airborne vibrations in crickets and feed-back control of calling song. In: Kalmring and Elsner (eds) Acoustic and vibrational communication in insects. Parey Verlag Berlin, pp 177–182

Dangles O, Ory N, Steinmann T, Christides JP, Casas J (2006) Spider's attack versus cricket's escape: velocity modes determine success. Anim Behav 72(3):603–610

Edgecomb RS, Robert D, Read MP, Hoy RR (1995) The tympanal hearing organ of a fly: phylogenetic analysis of its morphological origins. Cell Tiss Res 282:251–268

Farris HE, Oshinsky ML, Forrest TG, Hoy RR (2008) Auditory Sensitivity of an acoustic parasitoid (*Emblemasoma* sp., Sarcophagidae, Diptera) and the calling behavior of potential hosts. Brain Behav Evol 72:16–26

Fowler HG (1987) Field confirmation of the phonotaxis of *Euphasiopteryx depleta* (Diptera: Tachinidae) to calling males of *Scapteriscus vicinus* (Orthoptera: Gryllotalpidae). Fla Entomol 70:409–410

Fowler HG, Kochalka JN (1985) New record of *Euphasiopteryx depleta* (Diptera: Tachinidae) from Paraguay: attraction to broadcast calls of *Scapteriscus acletus* (Orthoptera: Gryllotalpidae). Fla Entomol 68:225–226

Fry SN, Bichsel M, Müller P, Robert D (2000) Three-dimensional tracking of flying insects using pan-tilt cameras. J Neurosci Methodol 101:59–67

Fullard JH, Yack JE (1993) The evolutionary biology of insect hearing. Trends Ecol Evol 8:248–252

Gnatzy W, Heusslein R (1986) Digger wasp against crickets. Naturwissenschaften 73:212–215

Gray DA, Banuelos C, Walker SE, Cade WH, Zuk M (2007) Behavioural specialization among populations of the acoustically orienting parasitoid fly *Ormia ochracea* utilizing different cricket species as hosts. Anim Behav 73:99–104

Hedwig, B Pollack GS (2008) Invertebrate auditory pathways. In: AI Basbaum, Kaneko A, GM Shepherd and G Westheimer editors: The Senses: A Comprehensive Reference, Vol. 3; Audition, P. Dallas and D. Oertel, San Diego Academic Press, pp 525–564

Hennig R, Franz A, Stumpner A (2004) Processing of auditory information in insects. Micros Res Tech 63(6):351–374

Köhler U, Lakes-Harlan R (2001) Auditory behaviour of a parasitoid fly (*Emblemasoma auditrix*, Sarcophagidae, Diptera). J Comp Physiol A 187:581–587

Lakes-Harlan R, Heller KG (1992) Ultrasound-sensitive ears in a parasitoid fly. Naturwissenschaften 79:224–226

Lakes-Harlan R, St·lting H, Stumpner A (1999) Convergent evolution of insect hearing organs from a preadaptive structure. Proc Roy Soc London. Series B: Biol.sci 266:1161–1167

Lakes-Harlan R, Jacobs K, Allen GR (2007) Comparison of auditory sense organs in parasitoid Tachinidae (Diptera) hosted by Tettigoniidae (Orthoptera) and homologous structures in a non-hearing Phoridae (Diptera). Zoomorphology 126:229–243

Lehmann GUC (2003) Review of biogeography, host range and evolution of acoustic hunting in Ormiini (Insecta, Diptera, Tachinidae), parasitoids of night-calling bushcrickets and crickets (Insecta, Orthoptera, Ensifera). Zool Anz 242:107–120

Lehmann GUC, Heller KG (1998) Bushcricket song structure and predation by the acoustically orienting parasitoid fly *Therobia leonidei* (Diptera: Tachinidae: Ormiini). Behav Ecol Sociobiol 43(4):239–245

Lehmann GUC, Heller K-G, Lehmann AW (2001) Male bushcrickets favoured by parasitoid flies when acoustically more attractive for conspecific females (Orthoptera: Phaneropteridae/ Diptera: Tachinidae). Entomol Gen 25:135–140

Léonide JC (1969) Recherches sur la biologie de divers diptères endoparasites d'Orthoptères. Memoires du Museum National d'histoire Naturelle serie A, Zoologie 53:1–246

Mason AC, Oshinsky ML, Hoy RR (2001) Hyperacute directional hearing in a microscale auditory system. Nature 410:686–690

Miles R, Robert D, Hoy R, (1995) Mechanically coupled ears for directional hearing in the parasitoid fly *Ormia ochracea*. J Acoust Soc Am 98(6):3059–3070

Miles RN, Hoy RR (2006) The development of a biologically-inspired directional microphone for hearing aids. Audiology and Neurootology 11(2):86–94

Miles RN, Su QT, Cui W, Shetye M, Degertekin FL, Bicen B, Garcia C, Jones SA, Hall NA (2009) A low-noise differential microphone inspired by the ears of the parasitoid fly *Ormia ochracea*. J Acoust Soc Am 125:2013–2026

Montealegre-Z F, Jonsson T, Robson-Brown K, Postles M, Robert D (2012) Convergent evolution between mammalian and insect audition. Science 338:968–971

Müller P, Robert D (2001) A shot in the dark: the silent quest of a free-flying phonotactic fly. JEB 204:1039–1052

Müller P, Robert D (2002) Death comes suddenly to the unprepared: singing crickets, call fragmentation, and parasitoid flies. Behav Ecol 13:598–606

Oshinsky ML, Hoy RR (2002) Physiology of the auditory afferents in an acoustic parasitoid fly. J Neurosci 22:7254–7263

Ramsauer N, Robert D (2000) Phonotactic orientation in a parasitoid fly. Naturwissenschaften 87:315–319

Robert D (2001) Innovative biomechanics for directional hearing in small flies. Biol Bull 200:190–194

Robert D (2005) Directional hearing in insects. In: Popper AN, Fay RR: Sound source localization. Springer Handbook of Auditory Research, pp 6–35

Robert D (2008). Sound localization in insects. In: Basbaum AL, Kaneko A, Shepherd GM and Westheimer G: The senses: a comprehensive reference, vol 3. Audition P Dallos and D Oertel. San Diego, Academic Press, pp 725–732

Robert D, Amoroso J, Hoy RR (1992) The evolutionary convergence of hearing in a parasitoid fly and its cricket host. Science 258:1135–1137

Robert D, Göpfert MC (2002) Novel schemes for hearing and acoustic orientation in insects. Curr Opin Neurobiol 12:715–720

Robert D, Read M, Hoy RR (1994) The tympanal hearing organ of the parasitoid fly *Ormia ochracea* (Diptera, Tachinidae, Ormiini). Cell Tiss Res 275:63–78

Robert D, Miles RN, Hoy RR (1996) Directional hearing by mechanical coupling in the parasitoid fly *Ormia ochracea*. J Comp Physiol A 179:29–44

Robert D, Hoy RR (2007) Auditory systems in insects. Invertebrate Neurobiology. Eds. R. Greenspan, G. North. Cold Spring Harbour Laboratory Press., In, pp 155–184

Robert D, Miles R, Hoy RR (1998) Tympanal mechanics in the parasitoid fly *Ormia ochracea*: intertympanal coupling during mechanical vibration. J Comp Physiol A 183:443–452

Robert D, Miles RN, Hoy RR (1999) Tympanal hearing in the sarcophagid parasitoid *Emblemasoma* sp.: the biomechanics of directional hearing. J Exp Biol 202:1865–1876

Römer H, Bailey W (1998) Strategies for hearing in noise: peripheral control over auditory sensitivity in the bushcricket *Sciarasaga quadrata* (Austrosaginae: Tettigoniidae). JEB 201:1023–1033

Römer H, Spickermann M, Bailey W (1998) Sensory basis for sound intensity discrimination in the bushcricket *Requena verticalis* (Tettigoniidae, Orthoptera). J Comp Physiol A 182:595–607

Schniederk·tter K, Lakes-Harlan R (2004) Infection behavior of a parasitoid fly, *Emblemasoma auditrix*, and its host cicada *Okanagana rimosa*. J Ins Sci 4(36):1–7

Shewell GE (1987) Sarcophagidae. In: Manual of nearctic diptera. Research Branch Agriculture Canada. Monogr 28, vol 2, pp 1159–1186

Soper R, Shewell G, Tyrrell D (1976) *Colcondamyia auditrix* nov. sp. (Diptera: Sarcophagidae), a parasite which is attracted by the mating song of its host, Okanagana rimosa (Homoptera: Cicadidae). The Can Entomol 108:61–68

Stölting H, Stumpner A, Lakes-Harlan R (2007) Morphology and physiology of the prosternal chordotonal organ of the sarcophagid fly *Sarcophaga bullata* (Parker). J Insect Physiol 53:444–454

Stumpner A, von Helversen D (2001) Evolution and function of auditory systems in insects. Naturwissenschaften 88:159–170

Stumpner A, Lakes-Harlan R (1996) Auditory interneurons in a hearing fly (*Therobia leonidei*, Ormiini, Tachinidae, Diptera). J Comp Physiol A 178:227–233

Stumpner A, Allen GR, Lakes-Harlan R (2007) Hearing and frequency dependence of auditory interneurons in the parasitoid fly *Homotrixa alleni* (Tachinidae: Ormiini). J Comp Physiol A 193:113–125

Tinghitella RM, Wang JM, Zuk M (2009) Preexisting behavior renders a mutation adaptive: flexibility in male phonotaxis behavior and the loss of singing ability in the field cricket. Behav Ecol 20:722–728

Tuck EJ, Windmill JFC and D. Robert D (2009) Hearing in tsetse flies? Morphology and mechanics of a putative auditory organ. Bull Entomol Res 99: 107–119

Walker TJ (1993) Phonotaxis in female *Ormia ochracea* (Diptera: Tachinidae), a parasitoid of field crickets. J Insect Behav 6:389–410

Walker TJ, Wineriter SA (1991) Hosts of a phonotactic parasitoid and levels of parasitism (Diptera: Tachinidae: *Ormia ochracea*). Fla Entomol 74:554–559

Wineriter SA, Walker TJ (1990) Rearing phonotactic parasitoid flies (Diptera: Tachinidae), Ormiini, Ormia spp.). Entomophaga 35:621–632

Yager DD (1999) Structure, development, and evolution of insect auditory systems. Microsc Res Tech 47(6):380–400

Zuk M, Simmons LW, Cupp L (1993) Calling characteristics of parasitized and unparasitized populations of the field cricket *Teleogryllus oceanicus*. Behav Ecol Sociobiol 33:339–343

Zuk M, Rotenberry JT, Tinghitella RM (2006) Silent night: adaptive disappearance of a sexual signal in a parasitized population of field crickets. Biol Letts 2:521–524

Chapter 5
Adaptive Sounds and Silences: Acoustic Anti-Predator Strategies in Insects

William E. Conner

Abstract There has been a recent resurgence of interest in the evolution of adaptive coloration and a new appreciation of the mechanisms, functions, and evolution of crypsis, aposematic coloration, and mimicry. I here apply these principles to the acoustic modality using insect examples and discuss adaptive silence, acoustic crypsis, stealth, acoustic aposematism, acoustic mimicry, and sonar jamming. My goal is to inspire students of bioacoustics to explore the full richness of the acoustic interactions between predator and prey in behavioral, physiological, and evolutionary contexts similar to those used by visual ecologists.

5.1 Introduction

Adaptive coloration encompasses a variety of concepts, including the beautiful and intricate exemplars of crypsis, warning coloration, and mimicry plus many other lesser-studied visual phenomena such as masquerade, countershading, and disruptive coloration. Early naturalists and insect lovers played a critical role in defining the behavioral significance of the color patterns of animals. Among them were Erasmus Darwin (1794), the father of Charles Darwin, who commented on the utility of concealing coloration for animals; Edward B. Poulton (1890), who coined the term "aposematism" for warning coloration; and Henry W. Bates and Fritz Müller, who defined Batesian mimicry (1862) and Müllerian mimicry (1878), respectively. Yet adaptive coloration is but a reflection of selective pressures imposed by predators that hunt visually as appreciated by a primarily visual audience—man. Predators that rely on sound rather than vision have selected for an entirely different suite of characteristics, and it is those adaptive sounds and

W. E. Conner (✉)
Department of Biology, Wake Forest University, Winston-Salem, NC 27106, USA
e-mail: conner@wfu.edu

B. Hedwig (ed.), *Insect Hearing and Acoustic Communication*,
Animal Signals and Communication 1, DOI: 10.1007/978-3-642-40462-7_5,

silences that I examine here. I admit to being inspired by a book titled *Avoiding Attack: the Evolutionary Ecology of Crypsis, Warning Signals and Mimicry* by Ruxton, Sherratt, and Speed (2004), and I happily recommend it to those looking for an insightful and a detailed treatment of adaptive coloration. The authors by their own admission, however, focus exclusively on the visual modality and rarely mention sound (but see Ruxton 2011). Given the number of predators that hunt orienting by sound that omission should be corrected. I hope to do that here.

Many predators use a combination of sensory modalities to hunt insects: vision, hearing, and olfaction are the major ones, with a bias toward vision in diurnal animals and toward hearing and olfaction in those that hunt at night. Acoustic specialists include nocturnal hunters such as owls, rodents, felids, canids which are all passive listeners, and bats which are both passive listeners and active echolocators. Insect counteradaptations often affect both types. The key predators are vertebrates which allow us to appreciate, at least in a rudimentary way, the problems associated with finding prey and interpreting acoustic prey signals. I will focus on the antipredator adaptations of terrestrial insects vis-à-vis these predators. I would be remiss not to include parasites as well. Whenever an insect uses the acoustic modality to communicate, it becomes vulnerable to parasites that eavesdrop on their sound signals (Zuk and Kolluru 1998; see Chap. 4 by Hedwig and Robert). Parasitism may prove equally important in shaping the acoustic communication systems of insects. Defenses can be divided into two categories: primary and secondary (Edmunds 1974). Primary defenses are those that prevent detection by the predator. Secondary defenses promote survival after detection. The primary defenses discussed below are adaptive silence, stealth as in the modern military sense of the word, and acoustic crypsis. Secondary defenses will include prevention of localization, acoustic aposematism, mimicry, and sonar jamming.

5.2 Avoiding Detection: Primary Defenses

Before proceeding, it would be wise to define a concept that is too frequently used without a precise definition: crypsis. To put it simply, crypsis is avoiding detection while remaining in plain sight. It includes a range of color and pattern strategies that prevent detection (Stevens and Merilaita 2009), including background matching, a general matching of the color, intensity, and pattern of the background; self-shadow concealment, the use of pattern to cancel telltale shadows that betray position; obliterative shading, similar shading that cancels other three-dimensional cues to recognition; disruptive coloration, markings that disrupt outlines and other recognition patterns; flicker-fusion camouflage, the blurring of fast-moving stripes to match general background color; distractive markings that draw attention away from recognition cues; transparency, allowing the background to show through decreasing visibility; and silvering, a high degree of reflectivity that prevents silhouettes in nondirectional light (for details see Stevens and Merilaita 2011). All of these examples describe visual phenomena. I would like to extend the definitions of Stevens and Merilaita (2009) to include all traits—visual, chemical, tactile, electric, and, here, acoustic cues—that minimize the probability of being detected when potentially detectable to an observer.

5.2.1 Adaptive Silence

In his massive tome on *Adaptive Coloration in Animals* H. B. Cott (1940) states that "cryptic silence is to the ear what cryptic appearance is to the eye." His examples refer to the stealth with which predators approach prey, but the concept can equally be applied to prey characteristics. Technically this is more akin to hiding in the visual modality, and some have referred to it as acoustic avoidance (Curio 1976). Certain moths, for instance, interrupt their sexual displays in the presence of acoustic predators (see Chap. 6 by Greenfield). Males of lesser waxmoth, *Achroia grisella,* advertise their presence to potential mates by producing ultrasonic 100 μs pulses at a rate of 100/s. Females are attracted by and walk toward those acoustic displays. In the presence of sounds mimicking, the search calls of gleaning bats with long pulse lengths >1 ms and lower pulse rates <30/s, the male moths shut down their acoustic advertisement calls (Spangler 1984; Greenfield and Baker 2003) to avoid predator detection. Similar adaptive silences have been reported in katydids and crickets (Spangler 1984; Belwood and Morris 1987; Faure and Hoy 2000; Bailey and Haythornthwaite 1998; ter Hofstede et al. 2010). The advantages and costs of such a strategy are straightforward; silence not only prevents detection by predators that hunt by listening, but also it prevents communication with conspecifics. Tradeoffs in such cases are inevitable.

"Whispering" moths illustrate another antidetection strategy (Nakano 2008). Male Asian corn borer moths, *Ostrinia furnicalis*, use specialized sex-specific scales on the forewings and mesothorax to produce very low intensity courtship songs with frequencies between 40 and 80 kHz. The male produces the sounds in the immediate vicinity of the female's ear. The implication is that the song provides a private communication channel between the male singer and the female listener, protecting the pair from conspecific competitors and predators. A variety of moths produce similar hushed songs, presenting new technical challenges to researchers studying moth courtship (Nakano 2009).

One of the most interesting examples of adaptive silence can be found in the cricket *Teleogryllus oceanicus* (Zuk et al. 2006, see Chap. 4 by Hedwig and Robert). In field crickets, the male stridulates by scissoring the wings and sliding a scraper on one wing across a file on the other. Females are attracted to the males' songs. *T. oceanicus*, an Australian and Pacific Island species, has been forced to alter that strategy. The species was introduced to the Hawaiian Islands in the late 1990s. On the islands of Oahu, Hawaii, and Kauai, it overlaps in distribution with the acoustically orienting parasitoid fly *Ormia ochracea*. Like the female cricket, the female fly is attracted to the male cricket's calling song. After locating the male cricket, the fly deposits her young on and in the vicinity of the male.

The voracious larvae burrow into the male, killing him as they develop. Within a remarkably rapid timeframe of about 20 generations intense selection by the fly has resulted in a morphological change in the wings of the male crickets on Kauai (Zuk et al. 2011). So-called flatwing males have wings similar in appearance to those of females. They have lost their file and scraper and are thereby silenced. These males produce no calling song and do not attract the parasitoid. It is not clear how the silent flatwing males now attract mates, although it has been suggested that they function as satellite males, waiting in the vicinity of calling males, and intercepting females as they move toward them (Cade 1980).

5.2.2 Stealth

In addition to shutting down acoustic signals that could attract gleaning predators, a prey insect may be able to dampen its echo signature to aerial hawking and gleaning bats by using a stealthlike mechanism analogous to the methods the military uses to reduce the echo signature of stealth aircraft to radar and submarines to sonar (Denny 2007). Several researchers have suggested that the scales on moth wings may decrease the amplitude of their telltale echoes by absorbing the echolocation cries of bats. This would give the moth a small but significant advantage in avoiding detection (Roeder 1967; Moss and Zagaeski 1994; Zeng et al. 2011). The scales of the nocturnal moths *Spilosoma niveus* and *Rhyparoides amurensis* (Erebidae) more than double the absorption factor of the wings for sounds between 40 and 60 kHz, resulting in a decrease of echo intensity of up to 2 dB over scaleless wings. This gives the moths with scales a 5–6 % increase in detection distance by the bat. Control butterfly wings did not show this effect, a result explained by the fact they butterflies are diurnal and not normally exposed to bat predation. While the mechanism of sound absorption remains to be determined, moth scales frequently have interstitial spaces between them and are typically covered with micropores and trabeculae reminiscent of manmade sound-absorbing materials (Zeng et al. 2011). I predict that researchers have just scratched the surface in this regard and many more stealth-like examples will be found that foil echolocating predators.

5.2.3 Acoustic Crypsis: Blending into Background Reflections

Tympanic organs (ears) sensitive to ultrasound have evolved multiple times in nocturnal Lepidoptera (Yack et al. 1999). In noctuoid family Erebidae, the tiger moths, the tympanic membranes are located on the thorax and are typically most sensitive to frequencies between 30 and 65 kHz, a frequency range that overlaps with the echolocation calls of bats. The tuning of individual moth species is often matched to the specific acoustic characteristics of the sympatric bat communities with which they have evolved (Fullard 1998 and references therein). Moth ears are elegantly simple in neurophysiological design with only 1–3 acoustically sensitive scolopidia attached to each tympanic membrane (Yack et al. 1999), yet they provide their bearers with critical information about their acoustic surroundings. Moth species capable of detecting bats take evasive action and have a clear survival advantage (up to 30–40 %) over species that lack ears (Corcoran and Conner 2012). Moths that lack ears have an alternate anti-bat strategy. As exemplified by earless ghost swift moths, *Hepialus humuli*, they can "become invisible" to echolocating bats by flying close (<0.5 m) above vegetation and effectively blending into the clutter of echoes emanating from the leaves and stems around them (Simmons et al. 1989; Rydell 1998). This is the closest analogy to crypsis in the visible world—a blending in with one's background. Many insects probably use this strategy. Unfortunately, the use of modern lighting defeats this strategy because insects are drawn up and out of their normal flights paths, no longer blend into the background and become easy prey for bats.

5.3 Preventing Attack Once Detected

5.3.1 Preventing Passive Localization

Signaling katydids, particularly in the tropics, are vulnerable to gleaning bats and arboreal vertebrates. Tree-dwelling katydids may decrease their vulnerability to predators by increasing the carrier frequency of their calls with commensurate decreases in their repetition rate (Morris et al. 1994). The increase in carrier frequency takes the calls beyond the upper frequency limit of vertebrate hearing and the decrease in repetition rate limits the opportunities for sound localization. These changes will, of course, incur costs, because the female katydids may have greater difficulty in locating singing males. In some cases, predation appears to have driven katydid species to use substrate-borne vibration by tremulation to render their communication signals more difficult for predators to detect and track (Belwood and Morris 1987; Belwood 1990).

5.3.2 Acoustic Aposematism

The antithesis of acoustic crypsis is advertising one's presence with gaudy sounds. In many cases that message warns the would-be predator of distasteful chemicals or other characteristics that make the prey less appetizing. Poulton (1890) coined the term "aposematism" from the Greek words *apo* (away from) and *semato* (sign or signal) to describe a warning signal. Among the first to fully appreciate the possibility of acoustic aposematism in insects was A. D. Blest of University College in London. In expeditions to Panama and Trinidad, he documented the acoustic displays of 183 species of tiger moths (Erebidae) and their relatives (Blest 1964) and then recorded the behavioral responses of *Cebus* monkeys and domestic fowl to many of them. He discovered a bewildering richness of visual and ultrasonic acoustic displays associated with this generally unpalatable group of insects. Forty years later, researchers are still trying to sort them out.

Many erebids are capable of producing ultrasound in the form of short, repetitive clicks. This sound is produced in response to tactile stimulation and/or the ultrasonic cries of insectivorous echolocating bats (Barber and Conner 2006 and references therein). The sound-producing organs of erebids are the tymbals, modified cuticular plates (episternites) on either side of the third segment of the thorax. Located ventro-laterally on the thorax, just above the third pair of legs, each tymbal is clearly visible as a translucent bubble often free of scales. The tymbals are driven by modified steering muscles beneath the cuticle. As the muscles contract and relax, the cuticle flexes in and out producing clicks (Fullard and Heller 1990). In many species, the anterior edge of the tymbal is adorned with small ridges called microtymbals. During the flexion and relaxation cycle (modulation cycle), the microtymbals produce microclicks which add complexity to the sounds produced by the entire tymbal. The resulting sounds register their peak intensity in the ultrasonic frequencies ranging from 30 to 75 kHz (Fullard and Fenton 1977; Barber and Conner 2006) and have sound pressure levels as high as 119 dB at 2 cm (Sanderford and Conner 1990). Sound production in erebids appears to have originated as a survival tactic against echolocating bats (Conner 1999).

Dorothy Dunning, in collaboration with Kenneth Roeder, was able to show that tymbal sounds had an effect on insectivorous bats (Dunning and Roeder 1965). Technologically unaided field observations suggested that the sounds were warning bats of their possessor's inherent bad taste (Acharya and Fenton 1992; Dunning et al. 1992; Dunning and Krüger 1995). Hristov and Conner (2005a,b) were the first to provide a solid experimental case for acoustic aposematism. They used high-speed videography, ultrasound recording, and a novel learning paradigm to dissect the role of sound in warning from other possible roles in startle and sonar jamming. Using moths that varied in both palatability and sound production they probed the ability of sound-naïve bats to respond to the anti-bat sounds produced by moths over a learning period of seven nights. They reasoned that if moth sounds warn bats of distasteful chemicals, naïve bats would capture the clicking moths at first and then learn to associate the clicks with bad taste and avoid clicking moths thereafter. In the absence of the distasteful chemicals, the sounds would be ineffectual over the course of the experiment. Their results supported both contentions: for naïve bats sounds were effective only when combined with bad taste (Fig. 5.1a and b).

Aposematism has been viewed as a gateway to increased diversification. Eleven of fourteen visually aposematic taxa examined were more speciose than nonaposematic sister groups (Przeczek et al. 2008). Not only does aposematism provide protection from predators, but also expands foraging niches and allows increased levels of specialization (Speed et al. 2009). Mathematical models predict that levels of chemical defenses should be correlated with signal strength, rendering

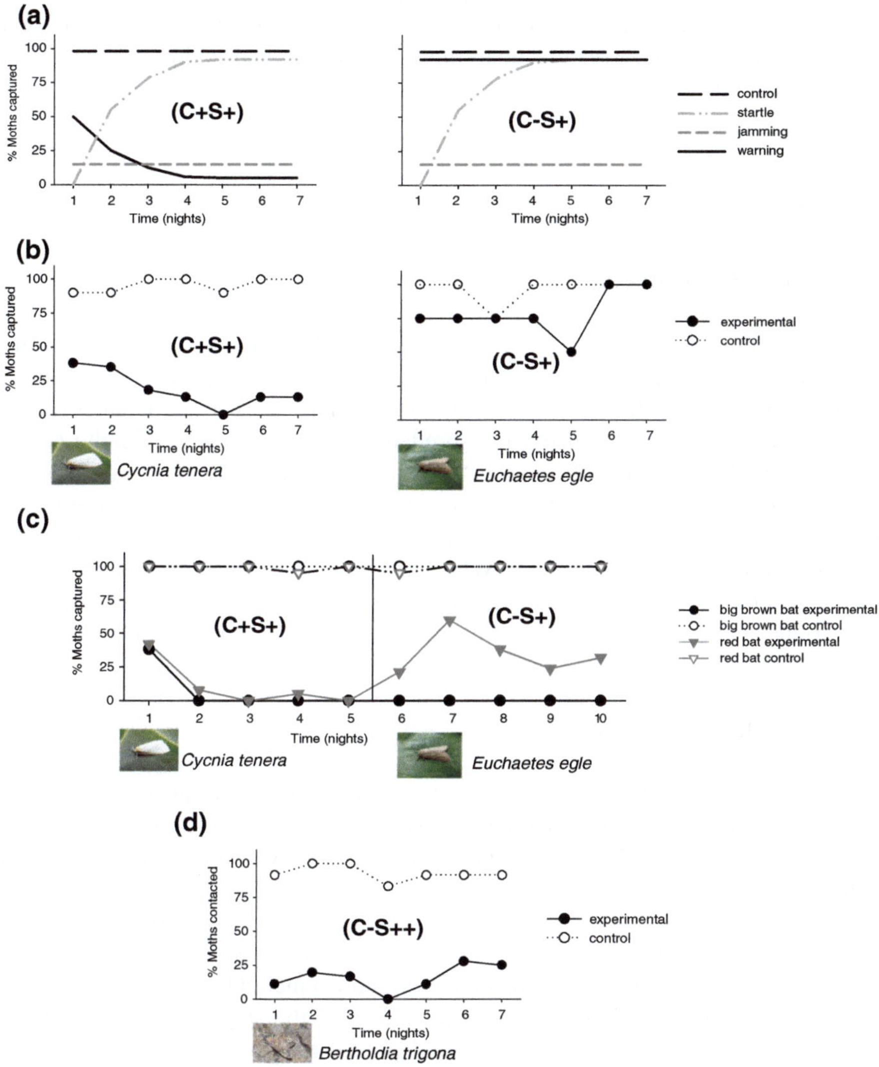

Fig. 5.1 Predicted and observed success of naïve bats attacking arctiid moths and moths used as controls over successive nights. **a** *Big brown* bat success attacking chemically defended (C+S+) and palatable (C–S+) clicking moths should vary according to three proposed moth click functions. **b** Acoustic aposematism experiment. Naïve *big brown* bat success attacking toxic, clicking *Cycnia tenera* and palatable, clicking *Euchaetes egle*. **c** Batesian mimicry experiment. *Red bats* and *big brown* bats were presented *C. tenera* for five nights, followed by *E. egle* for five nights. Intermediate *red bat* success on nights 6–10 reflect four bats that discovered the mimics and three bats that were fooled. **d** Jamming experiment, shows limited success of *big brown* bats attacking *Bertholdia trigona* over subsequent nights. Adapted from Conner and Corcoran (2012) with permission from the Annual Review of Entomology

aposematic signals honest (Speed et al. 2010). All of the studies mentioned here suggest testable hypotheses about the evolution of aposematic signals. In some cases they may be more readily quantified, controlled, and addressed within the acoustic modality.

5.3.3 Mimicry

5.3.3.1 Batesian Mimicry

If anti-bat sounds in general warn of distasteful chemistry, it seems likely that some palatable species will mimic the sound-producing species to mislead potential predators. This is the essence of Batesian mimicry. Barber and Conner (2007) were the first to test this concept in a predator/prey system involving the acoustic modality. They presented naïve bats with a distasteful clicking moth species (*Cycnia tenera*) for a five-day training session. Both species tested, the big brown bats, *Eptesicus fuscus*, and the red bats, *Lasiurus borealis*, learned to avoid the clicking moths after tasting a few. When the same bats were offered *Euchaetes egle*, a generally palatable species that produces similar sounds, the bats declined them. Batesian mimicry thus worked under these laboratory conditions (Fig. 5.1c).

5.3.3.2 Müllerian Mimicry

Barber and Conner (2007) also gathered evidence for Müllerian mimicry using a similar paradigm but with two unpalatable species. The results were similar to those described above. The bats wanted nothing to do with either species, which likely benefited from spreading the education of the predators over both species. Interestingly, the mimetic sounds need not be perfect replicas. The punishment associated with a bad choice may be sufficient to promote generalization of meaning; that is, if it clicks, it must be bad (Barber and Conner 2009).

These findings were important because they set the stage for a new view of acoustic communication between prey and predator, including the possibility of mimicry rings reminiscent of the stunningly mimetic visual patterns found in tropical butterflies. Moths may not be the only insects involved in such rings. David Yager and his colleagues showed that tiger beetles produce sounds when targeted by echolocating bats (Yager et al. 2000), and mimicry rings may extend even further. The key is, of course, to see if acoustic aposematism and mimicry work in the field. Experiments now in progress in North Carolina and at the Southwest Research Station in Cave Creek Canyon near Portal, Arizona, may answer these questions (Nick Dowdy, Wake Forest University, personal communication). High-resolution video, synchronized ultrasound recordings, and voice-recognition software for both bats and moths will make it possible to track both predators and prey noninvasively as they interact in nature.

5.3.4 Startle Signals

The sounds produced by the tiger moths may have additional functions, first referred to as deimatic (startling or frightening) signals (Edmunds 1974). Sudden and intense sounds stimulate the mammalian startle reflex (Hoy 1994) and trigger a moment of indecision on the part of a predator that opens a window of opportunity through which prey may escape. My students frequently drop cicadas when they are surprised by the insects' intense tymbalar sound. In this sense the sounds would function somewhat like the eyespots on the hind wings of a moth that can be suddenly exposed during an encounter with a predator (Blest 1957; Stevens et al. 2008). In addition, Ratcliffe and Fullard (2005) argued that signals of this type are more likely than neutral signals to be associated with aposematism. Through detailed analysis of the 3D interactions of big brown bats and the sound-producing tiger moth *Bertholdia trigona* Aaron Corcoran et al. (2011) have also gathered evidence that bats find the anti-bat sounds made by tiger moths at least initially startling. This function may be short-lived, as bats are thought to quickly habituate to the sounds (Miller 1991; Bates and Fenton 1990; Hristov and Conner 2005a). But as Ratcliffe and Fullard (2005) pointed out, we don't yet know the rates of exposure and habituation in the field, so the degree of importance of startle in the field remains an open question.

5.3.5 Sonar Jamming

Perhaps the most sophisticated function proposed for anti-bat moth sounds is that of sonar jamming. In theory, moth clicks could accomplish such jamming in three different ways. First, if moth clicks are sufficiently similar to returning prey echoes in spectral and temporal characteristics, bats might misperceive them as echoes from objects that do not exist, or "phantom targets" (Fullard et al. 1979; Fullard et al. 1994). Second, clicks that overlap or closely precede echoes may diminish a bat's precision in determining target range (Miller 1991; Surlykke and Miller 1985; Masters and Raver 1996; Tougaard et al. 1998, 2004). Finally, if clicks are sufficiently numerous and intense, they may mask the presence of echoes, rendering the target invisible (Troest and Møhl 1986; Møhl and Surlykke 1989). The three jamming hypotheses can be differentiated by what the bat perceives: multiple objects surrounding the moth for the phantom target hypothesis, a blurred target for the ranging interference hypothesis, and no target for the masking hypothesis (Corcoran et al. 2011).

For moth-produced sound to accomplish any of the three jamming mechanisms it should have certain characteristics, the most important being a high duty cycle. That is, the moth must produce clicks at a very high rate, filling the time after detection with sound to function as spurious targets, diminish ranging, or mask the presence of the target. Barber found such a moth in a survey of anti-bat sounds produced by moths in the cloud forest of western Ecuador (Barber and Conner

2007). Members of the genus *Bertholdia* (subfamily Arctiinae; family Erebidae) produce intense trains of clicks (duration = 0.28 ± 0.07 ms) at rates up to of 4500 clicks/s from each tymbal resulting in a duty cycle greater than 45 % (Corcoran et al. 2010). These are the hallmarks of a sonar jammer. Using a behavioral paradigm that pitted big brown bats against *Bertholida trigona,* Corcoran and his colleagues (2009) gathered the first behavioral evidence for sonar jamming (Fig. 5.1d). The bats had great difficulty contacting and capturing sound-producing *Bertholdia.* Analysis of the 3D flight tracts of the bats and moths (Corcoran et al. 2011) provided support for the hypothesis that the moths jam the ranging capability of the bats. The strategy has proven remarkably effective. It increases the probability of surviving bat attack by a factor of ten over survival of silenced individuals (Corcoran and Conner 2012). Interestingly, *Bertholdia* combines sonar jamming with physical evasion, simultaneously clicking and dropping, and thus presents a dual challenge to its attacker: to capture the moth the bat must concurrently overcome the errors in echo processing due to jamming and adjust its flight path to intercept the dropping moth (Ghose et al. 2006). In almost all cases, this proves to be too great a challenge and the bat is unable to capture the moth.

It is notable that the clicks of *Bertholdia* and of many erebid moths sum to a frequency-modulated signal. While the signals do not appear sufficient to mimic bat cries, moths may use a strategy akin to "sweep jamming" which allows jamming at multiple frequencies in rapid succession. This is an effective method of jamming radar and sonar in modern day human electronic warfare (Weik 1996; Denny 2007). Does sonar jamming have a visual analog? Abbott H. Thayer (1896, 1909), an artist with a profound appreciation of nature, first suggested that certain patterns of coloration would render animals difficult to track by making their trajectories and speeds ambiguous, a strategy referred to as dazzle coloration. The "dazzling" patterns are high-contrast stripes, bands, and zigzags that can create optical illusions when the prey is in motion (Stevens et al. 2008) and make it difficult for predators to effect capture. U.S. and British ships adopted similar patterns as camouflage in the Atlantic during World War II (Behrens 2009). According to the diminished ranging hypothesis, sonar-jamming moths are not difficult to detect, but their location is difficult to predict. This makes acoustic sonar jamming akin to dazzle coloration.

5.3.6 Anti-Bat Acoustic Diversity

Following in the footsteps of David Blest (1964), Barber and Conner (2007) and Corcoran et al. (2010) have recorded the anti-bat sounds of 30 species of tiger moths in the cloud forest of mainland Ecuador, in the piedmont region of North Carolina, and in Cave Creek Canyon in south-eastern Arizona. The moth sounds recorded in response to the hunting sequence of a typical FM bat vary from simple doublet clicks to complex frequency-modulated click trains (Conner and Corcoran 2012).

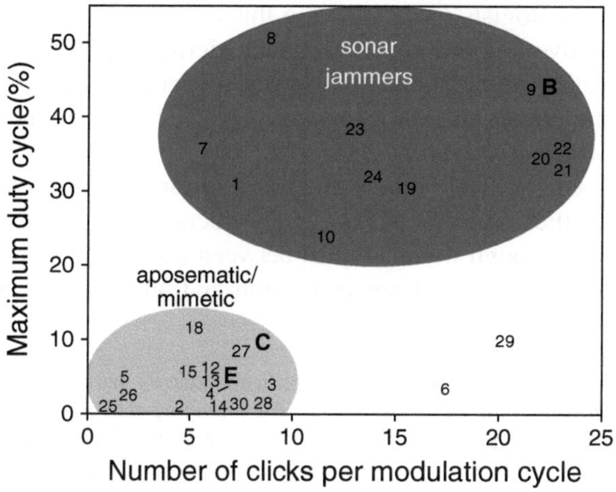

Fig. 5.2 Classification of moth anti-bat sounds plotted in acoustic space as defined by the maximum duty cycle (% of time occupied by sound) and modulation cycle complexity (# of clicks per modulation cycle (single flexion and relaxation of the tymbal organ). Each number represents a species. The sonar-jammers cluster (*dark grey*) contains *Bertholdia trigona* (B) and the aposematic/mimetic cluster (*light grey*) contains *Cycnia tenera* (C) and *Euchaetes egle* (E). Adapted from Conner and Corcoran (2012) with permission from the Annual Review of Entomology

One way to examine the function of the sounds produced by insects is to plot them in multidimensional sound space. Doing so allows us to appreciate the full array of sounds that insects produce and—equally important—the sounds that they do not produce. We recently did this exercise to characterize anti-bat sounds produced by moths. A principal components analysis (Corcoran et al. 2010) illustrated that anti-bat sounds form two clusters in acoustic space (Fig. 5.2). One cluster is composed of relatively simple click trains with low duty cycles, i.e., less that 20 % of the time contains clicks; the second is composed of complex click trains with high duty cycles, i.e., more than 20 % of the time contains clicks. Displaying the sounds this way allows visualization of the acoustic aposematism and mimicry strategies (Fig. 5.2, light grey cluster) and the sonar-jamming strategy (Fig. 5.2, dark grey cluster). The next step is to array the sounds on a phylogeny of the tiger moths to determine whether there are distinct aposematic, mimetic, and sonar-jamming clades. If so, which evolved from which? Or did the strategies evolve independently multiple times throughout the group?

5.4 Conclusion: Future Dimensions

This chapter is intended to stimulate thought and new research on adaptive sounds and silences. Simple words like crypsis, startle, aposematism, and mimicry mask exceedingly complex interactions between predator and prey. Bio-acousticians lag behind visual ecologists in deciphering this complexity. As has been the case in studies of adaptive coloration (Stevens and Merilaita 2011), the future should focus on the function of each sound and what it accomplishes for the prey—i.e., making range ambiguous, blending into the acoustic background, etc.—and the mechanism by which it exerts its effect, i.e., how it is perceived by the sensory systems of the predator. The latter is likely to be far more difficult than the former, but only together will they allow a full understanding of acoustic predator–prey interactions. Although my analogies between acoustic and visual signals are admittedly crude, they invite closer inspection and additional experimentation. I look forward to seeing the results.

Acknowledgments I would like to thank my graduate students, particularly Nickolay Hristov, Jesse Barber, Aaron Corcoran, and Nick Dowdy, for providing much of the bat/moth data that inspired this analysis. I would like to thank Wake Forest University for providing a Z. Smith Reynolds academic leave during which this chapter was written, and the Archbold Biological Station for providing an atmosphere conducive for carrying out the research and writing the manuscript. This material is based upon work supported by the National Science Foundation under Grant No. 0951160.

References

Acharya L, Fenton MB (1992) Echolocation behavior of verspertilionid bats (*Lasiurus cinereus and Lasiurus borealis*) attaching airborne targets including arctiid moths. Can J Zool 70:1292–1298

Bailey WF, Haythornthwaite S (1998) Risks of calling by the field cricket *Teleogryllus oceanicus*: potential predation by Australian long-eared bats. J Zool 244:505–513

Barber JR, Conner WE (2006) Tiger moth responses to a simulated bat attack: timing and duty cycle. J Exp Biol 209:2637–2650

Barber JR, Conner WE (2007) Acoustic mimicry in a predator-prey interaction. Proc Nat Acad Sc 104:9331–9334

Barber JR, Chadwell BA, Garrett N, Schmidt-French B, Conner WE (2009) Naïve bats discriminate arctiid moth warning sounds but generalize their aposematic meaning. J Exp Biol 212:2141–2148

Bates HW (1862) Contributions to an insect fauna of the Amazon valley. Lepidoptera: Heliconidae. Trans J Linn Soc 23:495–566

Bates DL, Fenton MB (1990) Aposematism or startle? Predators learn their responses to the defenses of prey. Can J Zool 68:49–52

Behrens RR (2009) Revisiting Abbott Thayer: non-scientific reflections about camouflage in art, war and zoology. Phil Trans Roy Soc B 364:497–501

Belwood JJ (1990) Anti-predator defences and ecology of neotropical forest katydids, especially the Pseudophyllinae. In: Bailey WJ, Rentz DCF (eds) The Tettigoniidae: biology, systematics, and evolution. Springer, New York

Belwood JJ, Morris GK (1987) Bat predation and its influence on calling behavior in neotropical katydids. Science 238:64–67

Blest AD (1957) The function of eyespot patterns in Lepidoptera. Behaviour 11:209–255

Blest AD (1964) Protective display and sound production in some New World arctiid and ctenuchid moths. Zoologica 49:161–181

Cade WH (1980) Alternative male reproductive behaviors. Fla Entomol 63:30–45

Conner WE (1999) "Un chant d'appel amoureux": acoustic communication in moths. J Exp Biol 202:1711–1723

Corcoran AJ, Barber JR, Conner WE (2009) Tiger moth jams bat sonar. Science 325:325–327

Corcoran AJ, Conner WE, Barber JR (2010) Anti-bat tiger moth sounds: form and function. Curr Zool 56:358–369

Corcoran AJ, Barber JR, Hristov NI, Conner WE (2011) How do tiger moths jam bat sonar? J Exp Biol 214:2416–2425

Corcoran AJ, Conner WE (2012) Sonar jamming in the field: effectiveness and behavior of a unique prey defense. J Exp Biol (in review)

Cott HB (1940) Adaptive coloration in animals. Methuen, London

Curio E (1976) The ethology of predation. Springer, Berlin

Darwin E (1794) Zoonomia. Johnson, London

Denny MW (2007) Bip, ping, and buzz: making sense of radar and sonar. Johns Hopkins University Press, Baltimore

Dunning DC, Roeder KD (1965) Moth sounds and insect-catching behavior in bats. Science 147:171–174

Dunning DC, Acharya L, Merriman CB, Ferro LD (1992) Interactions between bats and arctiid moths. Can J Biol 70:2218–2223

Dunning DC, Krüger M (1995) Aposematic sound in African moths. Biotropica 27:227–231

Edmunds M (1974) Defense in animals: a survey of anti-predator defenses. Longman, Harlow

Faure PA, Hoy RR (2000) The sounds of silence: cessation of singing and song pausing are ultrasound-induced acoustic startle behaviors in the katydid *Neoconocephalus ensiger* (Orthoptera: Tettigoniidae). J Comp Physiol A 186:129–142

Fullard JH (1998) The sensory coevolution of moths and bats. In: Hoy RR, Popper AN, Fay RR (eds) Comparative hearing: insects. Springer, New York

Fullard JH, Fenton B (1977) Acoustic and behavioural analyses of the sounds produced by some species of Nearctic Arctiidae (Lepidoptera). Can J Zool 55:1213–1224

Fullard JH, Heller B (1990) Functional organization of the arctiid moth tymbal (Insecta, Lepidoptera). J Morph 204:57–65

Fullard JH, Fenton MB, Simmons JA (1979) Jamming bat echolocation: the clicks of arctiid moths. Can J Zool 57:647–649

Fullard JH, Simmons JA, Saillant PA (1994) Jamming bat echolocation: the dogbane tiger moth *Cycnia tenera* times its clicks to the terminal attack calls of the big brown bat *Eptesicus fuscus*. J Exp Biol 194:285–298

Ghose K, Horiuchi TK, Krishnaprasad PS, Moss CF (2006) Echolocating bats use a nearly time-optimal strategy to intercept prey. PLOS Biol 4(5) e108 doi:10.1371/journal.pbio.0040108

Greenfield MD, Baker M (2003) Bat avoidance in non-aerial insects: the silence response of signaling males in an acoustic moth. Ethology 109:427–442

Hoy RR (1994) Ultrasonic startle in flying insects: some neuroethological and comparative aspects. Fortsch Zool 39:227–241

Hristov NI, Conner WE (2005a) Sound strategy: acoustic aposematism in the bat-tiger moth arms race. Naturwissenschaften 92:164–169

Hristov NI, Conner WE (2005b) Effectiveness of tiger moth (Lepidoptera, Arctiidae) chemical defenses against an insectivorous bat (*Eptesicus fuscus*). Chemoecology 15:105–113

Masters WM, Raver KA (1996) The degradation of distance discrimination in big brown bats (*Eptesicus fuscus*) caused by different interference signals. J Comp Physiol A 179:703–713

Miller LA (1991) Arctiid moth clicks can degrade the accuracy of range difference discrimination in echolocating big brown bats, *Eptesicus fuscus*. J Comp Physiol A 168:571–579

Møhl B, Surlykke A (1989) Detection of sonar signals in the presence of pulses of masking noise by the echolocating bat, *Eptesicus fuscus*. J Comp Physiol A 165:119–124

Morris GK, Mason AC, Wall P, Belwood JJ (1994) High ultrasonic and tremulation signals in neotropical katydids (Orthoptera: Tettigoniidae). J Zool 233:129–163

Moss CF, Zagaeski M (1994) Acoustic information available to bats using frequency-modulated sounds for the perception of insect prey. J Acoust Soc Amer 95:2745–2756
Müller F (1878) Über die Vortheile der Mimicry bei Schmetterlingen. Zool Anz 1:54–55
Nakano R, Skals N, Takanashi T, Surlykke A, Koike T, Yoshida K, Maruyama H, Tatsuki S, Ishikawa Y (2008) Moths produce extremely quite ultrasonic courtship songs by rubbing specialize scales. PNAS 105:11812–11817
Nakano R, Ishikawa Y, Taksuki S, Skals N, Surlykke A, Takanashi T (2009) Private ultrasonic whispering in moths. Commun Integrat Biol 2:123–126
Poulton EB (1890) The colours of animals: their meaning and use: especially considering in the case of insects, 2nd edn. Degan Paul, Trench Trübner & Co., London
Przeczek K, Mueller C, Vamosi SM (2008) The evolution of aposematism is accompanied by increased diversification. Integr Zool 3:149–156
Ratcliffe JM, Fullard JH (2005) The adaptive function of tiger moth clicks against echolocation bats: an experimental and synthetic approach. J Exp Biol 208:4689–4698
Roeder KD (1967) Nerve cells and insect behavior. Harvard University Press, Cambridge
Ruxton GD, Sherratt TN, Speed MP (2004) Avoiding attack: the evolutionary ecology of crypsis, warning signals and mimicry. Oxford University Press, Oxford
Ruxton GD (2011) Evidence for camouflage involving senses other than vision. In: Stevens M, Merilaita S (eds) Animal camouflage: mechanisms and function. Cambridge University Press, Cambridge, pp 330–350
Rydell J (1998) Bat defense in lekking ghost swifts (*Hepialus humuli*), a moth without ultrasonic hearing. Proc R Soc Lond B 265:1373–1376
Sanderford MV, Conner WE (1990) Courtship sounds of the Polka-dot Wasp Moth: *Syntomeida epilais*. Naturwiss 77:345–347
Simmons JA, Freedman EG, Stevenson SB, Chen L, Wohlgenant TJ (1989) Clutter interference and the integration time of echoes in the echolocating bat, *Eptesicus fuscus*. J Acoust Soc Am 86:1318–1332
Spangler H (1984) Silence as a defense against predatory bats in two species of calling insects. Southwest Nat 29:481–488
Speed MP, Brockhurst MA, Ruxton GD (2009) The dual benefits of aposematism: predator avoidance and enhanced resource collection. Evolution 64(6):1622–1633
Speed MP, Ruxton GD, Blount JD, Stephens PA (2010) Diversification of honest signals in a predator-prey system. Ecol Lett 13:744–753
Stevens M, Merilaita S (2009) Animal camouflage: current issues and new perspectives. Phil Trans R Soc B 364:423–427
Stevens M, Merilaita S (2011) Animal camouflage: mechanisms and function. Cambridge University Press, Cambridge
Stevens M, Yule DH, Ruxton GD (2008) Dazzle coloration and prey movement. Proc R Soc B 275:2639–2643
Steven M, Hardman CJ, Stubbins CL (2008) Conspicuousness, not eye mimicry makes 'eyespots' effective anti-predator signals. Behav Ecol 19:525–531
Surlykke A, Miller LA (1985) The influence of arctiid moth clicks on bat echolocation; jamming or warning? J Comp Physiol A 156:831–843
ter Hofstede HM, Kalko EKV, Fullard JH (2010) Auditory-based defence against gleaning bats in neotropical katydids (Orthoptera: Tettigoniidae). J Comp Physiol A 196:349–358
Thayer AH (1896) The law which underlies protective coloration. Auk 13:477–482
Thayer GH (1909) Concealing coloration in the animal kingdom: an exposition of the laws of disguise through color and pattern: Being a summary of Abbott H. Thayer's discoveries. Macmillan, New York
Tougaard J, Cassedy JH, Covey E (1998) Arctiid moths and bat echolocation: broad-band clicks interfere with neural responses to auditory stimuli in the nuclei of the lateral lemniscus of the big brown bat. J Comp Physiol A 182:203–215

Tougaard J, Miller LA, Simmons JA (2004) The role of arctiid moth clicks in defense against echolocating bats: interference with temporal processing. In: Thomas JA, Moss CF, Vater M (eds) Advances in the study of echolocation in bats and dolphins. University of Chicago Press, Chicago

Troest N, Møhl B (1986) The detection of phantom targets in noise by serotine bats; negative evidence for the coherent receiver. J Comp Physiol A 159:559–567

Weik MH (1996) Communications standard dictionary. 3rd edn. Chapman & hall, New york

Yack JE, Scudder GGE, Fullard JH (1999) Evolution of the metathoracic tympanal ear and its mesothoracic homologue in the Macrolepidoptera (Insecta). Zoomorphology 119:93–103

Yager DD, Cook AP, Pearson DL, Spangler HG (2000) A comparative study of ultrasound-triggered behavior in tiger beetles (Cicindelidae). J Zool Lond 251:355–368

Zeng J, Xiang N, Jiang L, Jones G, Zheng Y, Liu B, Zhang S (2011) Moth wing scales slightly increase absorbance of bat echolocation calls. PlosOne 6(11):e27190. doi:10.1371/journal.pone.0027190

Zuk M, Kolluru GR (1998) Exploitation of sexual signals by predators and parasitoids. Q Rev Biol 73:415–438

Zuk M, Rotenberry JT, Tinghitella RM (2006) Silent night: adaptive disappearance of a sexual signal in a parasitized population of field crickets. Biol Lett 2:521–525

Chapter 6
Acoustic Communication in the Nocturnal Lepidoptera

Michael D. Greenfield

Abstract Pair formation in moths typically involves pheromones, but some pyraloid and noctuoid species use sound in mating communication. The signals are generally ultrasound, broadcast by males, and function in courtship. Long-range advertisement songs also occur which exhibit high convergence with communication in other acoustic species such as orthopterans and anurans. Tympanal hearing with sensitivity to ultrasound in the context of bat avoidance behavior is widespread in the Lepidoptera, and phylogenetic inference indicates that such perception preceded the evolution of song. This sequence suggests that male song originated via the sensory bias mechanism, but the trajectory by which ancestral defensive behavior in females—negative responses to bat echolocation signals—may have evolved toward positive responses to male song remains unclear. Analyses of various species offer some insight to this improbable transition, and to the general process by which signals may evolve via the sensory bias mechanism.

6.1 Introduction

The acoustic world of Lepidoptera remained for humans largely unknown, and this for good reason: It takes place mostly in the middle- to high-ultrasound frequency range, well beyond our sensitivity range. Thus, the discovery and detailed study of acoustically communicating moths came about only with the use of electronic instruments sensitive to these sound frequencies. Such equipment was invented following the 1930s, and instruments that could be readily applied in the field were only available since the 1980s. But the application of such equipment

M. D. Greenfield (✉)
Institut de recherche sur la biologie de l'insecte (IRBI), CNRS UMR 7261, Parc de Grandmont, Université François Rabelais de Tours, 37200 Tours, France
e-mail: michael.greenfield@univ-tours.fr

B. Hedwig (ed.), *Insect Hearing and Acoustic Communication*,
Animal Signals and Communication 1, DOI: 10.1007/978-3-642-40462-7_6,

for studying acoustic animal species has accelerated in recent years, and we now know that certain moth species in several families use ultrasound for communication in much of the same way as Orthoptera and Cicadidae, as well as other acoustic species such as anurans, use audible sound. Scientists have known for a longer period that hearing, and specifically sensitivity to ultrasound, occurs in several major families of moths, that such perception is involved in the detection and evasion of predatory bats, and that some moth species also produce their own ultrasound emissions in the context of interacting with bats (see Chap. 5 by Conner). But if we define communication as an exchange of information between members of the same species, or at least as an event where both signaler and receiver benefit from the exchange (Hauser 1996; Greenfield 2002), then acoustic communication in moths is a rare behavior that has come to light only recently.

6.2 Lepidopteran Acoustics: Perception and Sound Production

Hearing, with specific sensitivity in the ultrasound frequency range, is very widespread in three major superfamilies of the Lepidoptera: Pyraloidea, Geometroidea, and Noctuoidea (Scoble 1992; see Minet 1983, 1985 for the Pyraloidea); these groups include over 90,000 species, comprising 55 % of lepidopteran species. The vast majority of pyraloid, geometroid, and noctuoid species hear, and those that do not may represent secondary loss of acoustic perception and tympanal organs. Ultrasound hearing is also found in the Drepanoidea, Thyridoidea, and Hedyloidea as well as in some Tineoidea, and Bombycoidea. In addition some butterflies (Papilionoidea) have tympanal hearing organs sensitive to sound in the audible frequency range. Considering both phylogeny and the homology of hearing organs it appears that hearing evolved independently 5–10 times in the Lepidoptera (Scoble 1992; Hoy and Robert 1996; Yack and Fullard 2000). The several groups have tympanal organs located in different body segments, including the abdomen (Pyraloidea, Geometroidea, Tineoidea, Drepanoidea), thorax (Noctuoidea), wing bases (Thyridoidea, Hedyloidea, Papilionoidea), and even the proboscis (Bombycoidea). A reanalysis of relationships among the lepidopteran superfamilies based on an extensive molecular dataset provides an improved phylogeny (Mutanen et al. 2010), and it may resolve previous uncertainty over clades and thereby allow a more definitive assessment of the evolution of hearing in the order.

Phylogeny combined with historical biogeography and the evidence from fossil Lepidoptera indicate that the evolutionary origins of lepidopteran hearing, with the possible exception of that in the Tineoidea (Davis 1998), occurred not earlier than 60 mya (Yack and Fullard 2000). This date suggests strongly that lepidopteran hearing originated in response to predatory bats hunting aerial insects via echolocation. The predation pressure from bats may have been critical in maintaining hearing in the Lepidoptera, as hearing and tympanal organs are reduced or even lost in pyraloid, geometroid, and noctuoid species inhabiting regions free of insectivorous bats (Fullard 1994). Similarly, species that have escaped bat predation temporally by activity during times of the day or year when bats are not active also show hearing loss (Fullard et al. 1997; Surlykke et al. 1998). It has been suggested that the Papilionoidea (butterflies), which generally do not hear, at least not ultrasound, have escaped bat predation by evolving from an ancestral state of nocturnal behavior to their present daytime activity (Yack and Fullard 2000). Some anatomical features of butterflies can be interpreted as vestiges of ancestral hearing organs, and interestingly their sister group, the Hedyloidea, are nocturnal and do hear ultrasound. The anatomical and neural apparatus necessary for hearing may come with a cost, and when hearing is no longer needed the investment devoted to these features is reduced (cf. Yager 1999).

Although hearing is widespread in moths, sound production in the context of mating communication occurs only in isolated genera and species scattered among various clades within the Pyraloidea and Noctuoidea (Conner 1999). Some of these acoustically communicating species also broadcast sounds in the context of interacting with bats (cf. Dunning and Roeder 1965), and in several cases the same sounds serve both as mating signals and during interactions with predators (Conner and Corcoran 2012). However, in most acoustically communicating moths the sounds produced serve only in mating. That is, these moths may typically hear and respond evasively to the echolocation signals of bats, but echolocation signals do not elicit any sounds from the moths that might serve in jamming, warning, or startling bat predators. Whereas tympanal hearing organs within the Pyraloidea and Noctuoidea appear to be homologous within each superfamily, sound producing structures are situated on many different locations (Greenfield 2002; Nakano et al. 2009b). These points have major implications for the evolution of acoustic communication in the Lepidoptera and in general.

6.3 Acoustic Communication in the Lepidoptera: History of a Discovery

The death's head hawkmoth (*Acherontia atropos*; Sphingidae: Sphinginae), in which the adults enter colonies of bees to imbibe honey, was perhaps the first acoustic lepidopteran whose sounds drew human attention (Kitching 2003). Adult *A. atropos* emit sounds by expelling air through the mouthparts when disturbed, and it had been proposed that these emissions mimic the 'piping' sounds of the queen and thus inhibit worker honeybees from attacking the moths while they are feeding (Kettlewell 1978), a suggestion however questioned by Moritz et al. (1991). During the nineteenth century several reports described sounds emitted by various species of moths and butterflies, and the presence of 'tympaniform' organs, assumed to be involved in hearing, was noted as well (Laboulbène 1864; review in McLachlan 1877). Some of the sounds reported occur during interactions with bats, and they are audible because of their broadband nature. It was even suggested that lepidopteran acoustic signaling may be 'more general than commonly believed' (McLachlan 1877).

The first definitive report of sound used in mating communication in the Lepidoptera was made in the Lesser waxmoth (*Achroia grisella*; Pyralidae: Galleriinae), a species with distinctive pair formation wherein males attract females up to 1 m distant with a high-frequency song (Spangler et al. 1984). In most species of moths females emit an advertisement pheromone that attracts males over relatively long distances (Greenfield 1981). Males may also emit a close-range courtship pheromone that influences female acceptance, but the basic scheme of signaling females and searching males is the rule in moths. *A. grisella* is an exception to this rule, as males attract females with what appeared to be some type of mechanical stimulation (Dahm et al. 1971). Males generate this stimulation by continuously fanning their wings while remaining stationary on the substrate (Kunikë 1930; Greenfield and Coffelt 1983). Neighboring males may occasionally run toward and butt a wing-fanning individual, which results in a brief scuffle, repositioning, and resumption of wing-fanning by one or both of the males. A female will move among the males and eventually run toward a given individual, her trajectory becoming rather direct once she begins the final approach. Although females too may fan their wings while approaching, males do not move or even turn toward the approaching female until she makes physical contact, at which time a brief courtship and mating ensue. These observations, combined with knowledge that the moths can hear very high-frequency ultrasound (Spangler and Takessian 1983), suggested that *A. grisella* was an acoustic species and that the male wing-fanning was generating an acoustic mating signal. Sound recordings then showed that while females are mute, male wing-fanning caused a pair of tymbal organs situated on the tegulae, small sclerites at the bases of the forewings, to resonate and to produce rhythmic trains of 100 kHz pulses. Female muteness reflects their lack of tymbals. Behavioral tests demonstrated that wing-fanning males whose tegulae were ablated did not attract females, but broadcasts of synthetic male song from a loudspeaker were as attractive to females as wing-fanning males (Spangler et al. 1984). The findings on *A. grisella* demonstrated that moths do use sound for purposes other than sending signals to predators or hosts, and that when they do communicate acoustically the broadcasts are likely to be in the ultrasound frequency range. Similar findings were later made in other moth species, revealing male songs involved in long-range advertisement as well as close-range courtship.

6.4 Acoustic Communication in the Lepidoptera: Phylogenetic Survey

To date, acoustic communication has been reported in the Pyraloidea (Pyralidae, Crambidae), Noctuoidea (Noctuidae, Arctiidae), and Papilionoidea (Nymphalidae). This section provides an overview of which groups and species broadcast mating songs and the manner in which these songs are produced and function. Systematic treatment follows the phylogeny of the Lepidoptera presented in the Tree of Life Web Project 2010 (www.tolweb.org; Fig. 6.1).

6.4.1 Pyraloidea

Several species of galleriine pyralid moths produce male courtship songs that influence female response, including *Aphomia sociella* (Figs. 6.2a and 6.3a; Kindl et al. 2011) and *Eldana saccharina* (Bennett et al. 1991), and possibly *Galleria mellonella* (Spangler 1985, 1986). Only *Achroia grisella* is known to produce a long-range male advertisement song (Spangler et al. 1984). Songs that give the appearance of functioning in courtship but which have not been tested experimentally are observed in *Corcyra cephalonica* (Spangler 1987) and *Achroia innotata* (personal observations), a congener of *A. grisella*. In all of the above species males sing while stationary and fanning their wings, which causes tymbals on the tegulae to resonate. *Achroia*, *Aphomia*, and *Galleria* are symbionts of social bees, in whose colonies the moth larvae develop. However, no observations indicate that male song in these galleriine species has any influence on bee behavior. In both *A. sociella* and *G. mellonella* males attract females with long-range sex pheromones, and the male song probably acts only in close-range courtship. *A. grisella* males also emit an odor that originates from glands on the forewings, but the role of

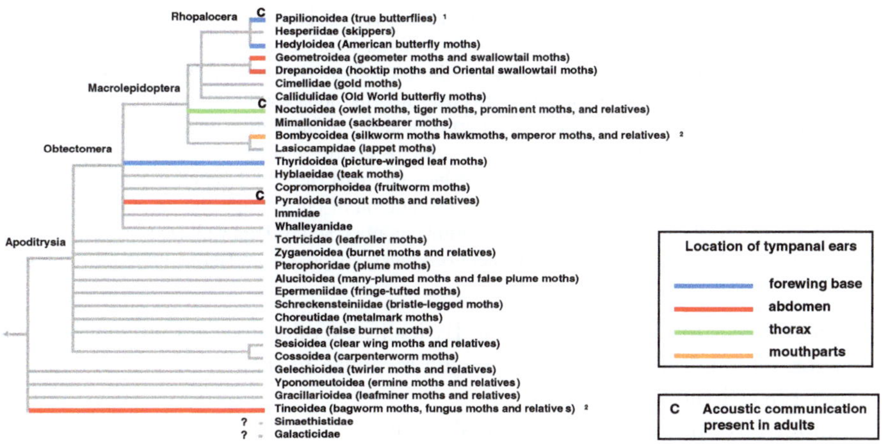

Fig. 6.1 Phylogeny of the ditrysian Lepidoptera showing distribution of tympanal ears, classified by anatomical location, and acoustic mating communication, denoted by 'C'. Unless indicated by a superscript, tympanal ears are found in the majority of species in the clades represented by colored branches. 1 Tympanal ears in some groups of Papilionoidea are sensitive to audible sound, 2 Tympanal ears in some groups of Bombycoidea, and Tineoidea are ultrasound sensitive

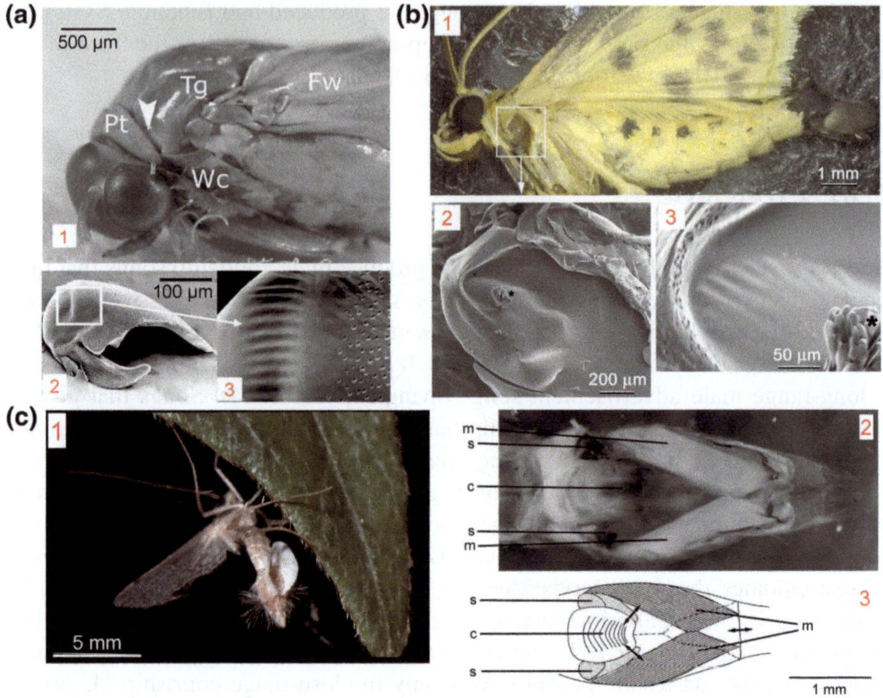

Fig. 6.2 Sound-producing structures in male pyraloid moths. **a** *Aphomia sociella*: 1. general situation of the tymbal (*arrow*) on the tegula (*Tg*), to the left of the forewing base (*Fw*) and to the right of the patagium (*Pt*), tegular wing coupler (*Wc*); 2. tymbal located on the male tegula; 3. detail of striations on the male tymbal (adapted from Kindl et al. 2011), **b** *Conogethes punctiferalis*: 1. male tymbal situated laterally on the mesothorax, indicated by box (adapted from Nakano et al. 2012b); 2 and 3. scanning electron micrographs showing detail of membrane on male tymbal; black asterisk, indicating a group of scales that had fallen on the membrane, serves to relate the higher magnification Figs. 6.2–6.3 (photos courtesy of Ryo Nakano), **c** *Symmoracma minoralis*: 1. singing male (photo courtesy of Klaus-Gerhard Heller); 2. anterior of last abdominal segment of male, dorsal view; 3. schematic representation of photograph shown in part 2, indicating cavity (*c*) with ribbed bottom serving as sound chamber, sclerotized wall (*s*) of cavity, muscles (*m*), and proposed sound-producing movement (*arrows*) (adapted from Heller and Krahe 1994)

this scent in pair formation is not clear. By itself, the odor does not influence female behavior, although it may enhance female response to male song (Greenfield and Coffelt 1983; Spangler et al. 1984).

Among phycitine pyralid moths, male songs have been recorded in several species (*Ephestia cautella*, *E. kuehniella*, *Plodia interpunctella*). These moths exhibit typical pair formation, long-range female pheromones, and short-range male courtship pheromones, and it is suggested that the songs act in concert with the latter (Trematerra and Pavan 1995). The sound-producing structures in phycitine moths have not been described.

Loud male advertisement songs have been recorded in several species of crambid moths (Figs. 6.2b, c and 6.3b, c). In *Syntonarcha iriastis* (Crambidae:

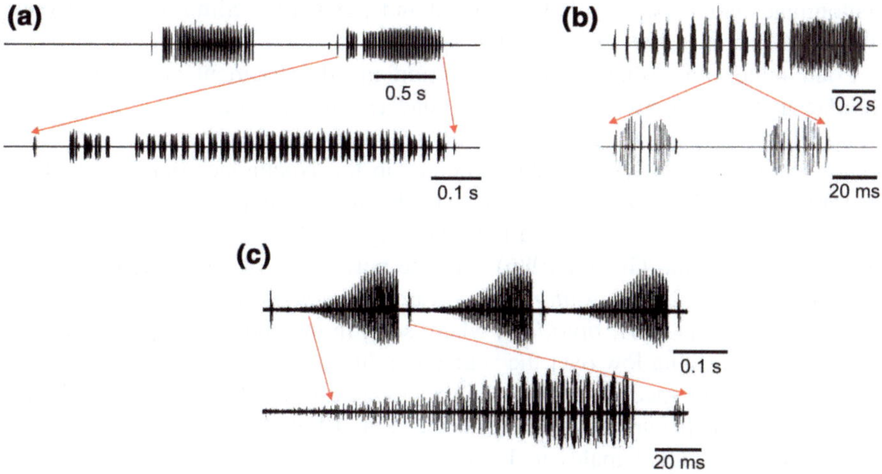

Fig. 6.3 Male songs in pyraloid moths. Oscillograms show representative segments of ultrasound song in: **a** *Aphomia sociella* (adapted from Kindl et al. 2011), **b** *Conogethes punctiferalis* (adapted from Nakano et al. 2012a), **c** *Symmoracma minoralis* (adapted from Heller and Krahe 1994), **a** and **b** are courtship songs, **c** may be an advertisement song, but its function is not yet confirmed

Odontiinae) and *Symmoracma minoralis* (Crambidae: Spilomelinae) songs are produced by structures on the genitalia; in the former the structure is stridulatory (Gwynne and Edwards 1986) whereas in the latter it is a tymbal organ (Heller and Krahe 1994). In *Conogethes punctiferalis* (Crambidae: Spilomelinae), sound is produced by a tymbal organ situated on the mesothorax, and the role of male song in courtship has been confirmed (Nakano et al. 2012a, b).

Songs in the above species of Galleriinae and Crambidae are all broadcast at amplitudes $\geq$80 dB SPL at 1 cm from the singing male (Nakano et al. 2009b). In other species of Crambidae, as well as Noctuidae and Arctiidae, males have been found to whisper 'soft songs' at much lower amplitudes, <60 dB SPL at 1 cm (Nakano et al. 2009a). These songs may be produced by stridulation with specialized scales on the wings and thorax (Nakano et al. 2008). The extremely low amplitude, nonetheless detectable by the female, may prevent eavesdropping predators or rival males from localizing a courting pair. In *Ostrinia* spp. (Crambidae), the role of quiet songs in courtship has been confirmed by testing males whose specialized scales were ablated (Takanashi et al. 2010).

6.4.2 Noctuoidea

Acoustic communication has been reported in four noctuoid families, Nolidae, Lymantriidae, Arctiidae, and Noctuidae. Mating songs are best known in the Arctiidae, where they are reported in the Arctiinae, the Lithosiinae, and the

Syntominae and appear to have evolved independently (Simmons and Conner 1996; Weller et al. 1999; DaCosta et al. 2006).

Loud male songs that probably function in advertisement are found in the Nolidae and the Noctuidae. In the Nolidae (*Pseudoips prasinana*; *Bena bicolorana*), tymbals located ventrally on the abdomen serve as the sound producing organs (Skals and Surlykke 1999), whereas in the Noctuidae sound is produced by forewing tymbals (*Amyna natalis*) (Heller and Achmann 1995), stridulation between the metathoracic leg and the hindwing (*Rileyana* (formerly *Thecophora) fovea*) (Surlykke and Gogala 1986), and the percussion of castanet structures on the forewings (*Hecatesia exultans*, *Hecatesia thyridon*) (Bailey 1978). Males sing while at rest except in *H. thyridion*, where song in the audible frequency range is produced while flying low over the vegetation during the day (Alcock et al. 1989). In both *Hecatesia* species, males exhibit agonistic interactions and defend small territories at which females arrive (Alcock et al. 1989; Alcock and Bailey 1995). The attraction of local males to loudspeakers broadcasting male song indicates that sound plays an important role in these interactions and territorial defense.

In the Arctiidae acoustic communication is often related to sounds produced in the context of interacting with bats. Male *Euchaetes egle* and *Euchaetes bolteri* produce species-specific songs with thoracic tymbal organs when courting females. These songs are acoustically identical to the sounds that both sexes emit for warning or jamming bats, and there is no sexual dimorphism in tymbal structure (Simmons and Conner 1996). In other species (e.g. *Phoenicoprocta capistrata*) the tymbal structure and the sounds differ between the sexes (Rodriguez-Loeches et al. 2009). In *Empyreuma affinis* and *Syntomeida epilais* both sexes sing in a duetting fashion (Sanderford and Conner 1995, 1998), while in *Pyrrharctia isabella* it is only the female who sings, which she does in response to the male courtship pheromone (Krasnoff and Yager 1988). The female acoustic reply in *P. isabella* is identical to the sound she emits when disturbed. Male songs are necessary for courtship success in some species, but in others (e.g. *Cycnia tenera*) they may be critical only in the absence of male courtship pheromone (Conner 1987). Sanderford (2009) presents a more comprehensive survey of acoustic communication in the Arctiidae.

In the Lymantriidae soft songs are reported in several species and may be similar to those described in the previous section except that they are produced by the female, not the male, during courtship (Rowland et al. 2011). The mechanism of sound production is unknown.

6.4.3 Papilionoidea

Several species of *Hamadryas* butterflies (Nymphalidae) produce audible clicks by percussion of veins in the forewings. These sounds are generally emitted by both sexes in a defensive context. Males also click during agonistic interactions with neighbors, and in at least one species the sounds represent a courtship song (Monge-Najera et al. 1998).

6.5 Physiology and Behavior of a Model Acoustic Species, *Achroia grisella*

6.5.1 *Mating Communication*

The galleriine pyralid moth *Achroia grisella* (Fig. 6.4a) has served as a model species in the study of sexual selection over the past 20 years. I present this example to illustrate how sound can function in mating communication in the Lepidoptera, and also to infer the evolution of acoustic communication in the order.

A. grisella adults normally remain in the immediate vicinity of honeybee (*Apis mellifera*) colonies, and signaling and mating behavior take place there (Greenfield and Coffelt 1983). Males begin singing at dusk and continue more or less uninterruptedly for 6–10 h. This activity starts on the evening following adult eclosion and is repeated each night over 7–10 d until senescence and death. The male song is a series of brief ultrasound pulses (~100 μs, 70–130 kHz) delivered at a regular rhythm ranging from 80–100 pulses·s^{-1} among individuals within a population (Jang and Greenfield 1996; Jang et al. 1997; Fig. 6.4d). The pulse rhythm increases linearly with temperature (Greenfield and Medlock 2007) and reflects the rate of wing beating, which generates two pulses during each complete cycle of wing movement. A given pulse represents the strongly damped resonance of the tymbals on the tegulae (Fig. 6.4b, c), which are knocked by the wing bases during the upstroke and again during the downstroke. Closer analysis reveals that most

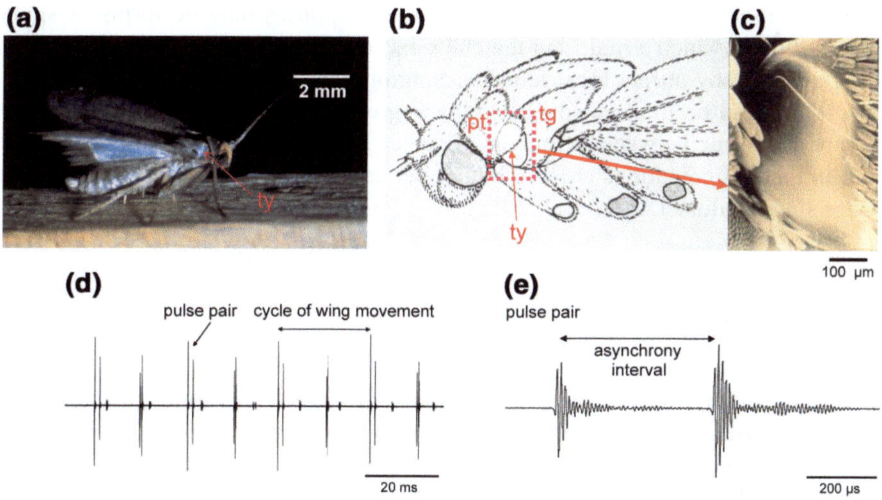

Fig. 6.4 Sound production and ultrasound advertisement song in male *Achroia grisella*. **a** singing male with sound producing tymbal (*ty*) shown, **b** schematic representation of tymbal (*ty*) situated on the tegula (*tg*), posterior and ventral to the patagium (*pt*), **c** scanning electron micrograph showing tymbal, the scale-less frontal area of the tegula (adapted from Spangler et al. 1984)

resonances are actually 'double pulses' (Fig. 6.4e) in which a brief (~100–500 μs) delay, termed the 'asynchrony interval', separates two individual pulses. The delay arises because movement of the left and right wings is not perfectly synchronous, and the two sides generate two non-overlapping pulses.

Female response to and preference for males is strongly influenced by the acoustic characters of the male song. Females move toward songs that are delivered at a faster pulse-pair rate, that include longer pulses and pulses with a higher peak amplitude, and whose pulse pairs encompass a longer asynchrony interval (Jang and Greenfield 1996; Limousin and Greenfield 2009; Alem and Greenfield 2010). Most of these preferences represent choice for higher acoustic energy or power and may be based on greater afferent activity. There is no evidence that females evaluate the carrier frequency of male song, which is not surprising given that the *A. grisella* ear has only 4 afferent neurons, each attached to the tympanum at the same central location. The tympanum is broadly tuned between 40 kHz and at least 100 kHz, and its peak sensitivity, between 90 and 100 kHz, matches the dominant frequencies of male song (Rodriguez et al. 2005). Females also exhibit a threshold response to the pulse-pair rate of male song (Brandt et al. 2005) and will not move toward a single song stimulus if it is delivered below a minimum rate ranging from 20–80 pulse pairs·s^{-1}. Behavioral tests with live males demonstrated that females evaluate the overall attractiveness of song based on a multivariate function of the several signal characters (Jang and Greenfield 1998), which may vary among individuals (Jang and Greenfield 2000) and between populations (Zhou et al. 2008).

Male song also evokes several male responses in *A. grisella*: Neighboring singers stimulate a male to initiate song at the beginning of the night, and a particularly close neighbor can elicit a 4–10 % increase in pulse-pair rate for 15–20 min during a singing bout (Jia et al. 2001). These responses may be forms of signal competition in which a male can match or exceed the quality of a neighbor's signaling and thereby attract local females. Singing males often gather in small clusters, which can be interpreted as lekking aggregations, and agonistic interactions commonly occur between neighbors (Cremer and Greenfield 1998). Males who sing in these aggregations enjoy a higher per capita attractiveness to females than males singing alone (Alem et al. 2011).

6.5.2 Sexual Selection

Male *A. grisella* who broadcast attractive songs expend more energy than males who broadcast songs of inferior attractiveness (Reinhold et al. 1998). Song attractiveness, as determined by the signal characters influencing female preference, is phenotypically and genetically correlated with a male's body mass at adult eclosion, the duration of his nightly singing period, and his adult longevity (Brandt and Greenfield 2004). That is, song attractiveness is 'condition' dependent, and in the laboratory there is no evidence of a trade-off between sexual attractiveness and survival.

Quantitative genetic analyses showed that the various male signal characters are heritable traits (Collins et al. 1999; Brandt and Greenfield 2004; Zhou et al. 2011). Similar findings, albeit based on fewer data, were obtained for the female preference function (Jang and Greenfield 2000) and response threshold (Rodriguez and Greenfield 2003). But contrary to some expectations from sexual selection theory, no genetic covariance has been found between the male signal and female preference and response traits in *A. grisella* (Zhou et al. 2011; Limousin 2011; Limousin et al. 2012). Both male signaling and female preference are influenced by several quantitative trait loci (QTL) that are distributed among the 29 chromosomes. One QTL that is specifically associated with the pulse-pair rate of male song exerts a very major influence on this song character, but none of the song QTL are located on the same chromosomes as the preference QTL. The findings on the number and strength of QTL influencing song are similar to those in Hawaiian cricket species (Shaw et al. 2007).

6.5.3 On the Bat-Moth Interaction

In *A. grisella* behavioral responses to predatory bats and their echolocation signals occur in flight and also on the substrate, where males exhibit very different behavior from females (see Chap. 5 by Conner). Flying *A. grisella* cease beating their wings and drop toward the ground upon hearing echolocation signals (Rodriguez and Greenfield 2004). This response can be elicited by a single pulse of ultrasound (40–100 kHz) that exceeds a minimum duration of 1 ms and amplitude of 70 dB. Response latencies decrease as the pulse lengthens, the amplitude increases, or the pulses are delivered more rapidly and may be as short as 50 ms.

Because a considerable proportion of *A. grisella* activity takes place on the substrate, we asked whether the moths also respond to echolocation signals in this situation. Whereas foraging bats are not likely to enter wooden honeybee hives and prey on the moths, it is critical to recognize that these enclosed hive boxes have only been employed in apiculture since the middle of the nineteenth century. Moreover, much signaling and mating behavior in *A. grisella* takes place in the vicinity of the honeybee colony rather than within it. Thus, *A. grisella* would have been exposed to bat predation, particularly from bat species that glean their insect prey from the substrate, for most of their evolutionary history, and they remain exposed today. We found two basic responses of *A. grisella* to echolocation signals while on the ground: Singing males become temporarily silent (referred to as a 'silence response'), and females who are running, as during their approach to a singing male, cease movement (referred to as a 'stationary response'; Greenfield and Weber 2000; Greenfield and Baker 2003). Male and female responses are elicited by ultrasound pulses that exceed a minimum of 75 dB SPL and a duration of 1 ms but are delivered at a relatively slow rate, <30 pulses$\cdot$s^{-1}. This last criterion reflects the slow pulse rate of bat echolocation signals broadcast during the 'search phase' of predation. It is inferred that *A. grisella* discriminate singing male conspecifics from searching bats on the basis of pulse rate rather than pulse length (Greenfield and Weber 2000). Whereas bat echolocation signals are normally much longer than the pulses in *A. grisella* male song, female *A. grisella* will respond positively to artificially lengthened pulses as if they are super-normal stimuli (Fig. 6.5).

Tests in a flight room in which *A. grisella* were exposed to live bats—*Rhinolophus ferromequinum* (greater horseshoe bat), a species that both forages aerially and gleans prey from the substrate and has no aversion to eating *A. grisella* in captivity—revealed both male silence and female stationary responses to bat echolocation signals (Alem et al. 2011). More importantly, the bats were attracted toward singing male *A. grisella*. Thus, the silence response in males is probably an adaptation that reduces the risk of exposure to bat predation. Because moths moving on substrates generate inadvertent sounds that may be perceived by gleaning bats, the stationary response in females is probably a defensive adaptation as well. Life history and the 'acoustic scene' influence the balance between sexual advertisement and defensive behavior in *A. grisella*: Males and females are less likely to exhibit silence and stationary responses as they age and the probability of future mating opportunities declines (Lafaille et al. 2010). The silence response is also reduced when males join lekking aggregations because neighbors' songs mask predator signals. Moreover, the pressure of signal competition in a lek may 'motivate' males to ignore predation risks (Brunel-Pons et al. 2011).

The dual function of hearing in *A. grisella*, and in moths in general, invites questions on the genetics of sexual and defensive behavior. Do sexual and

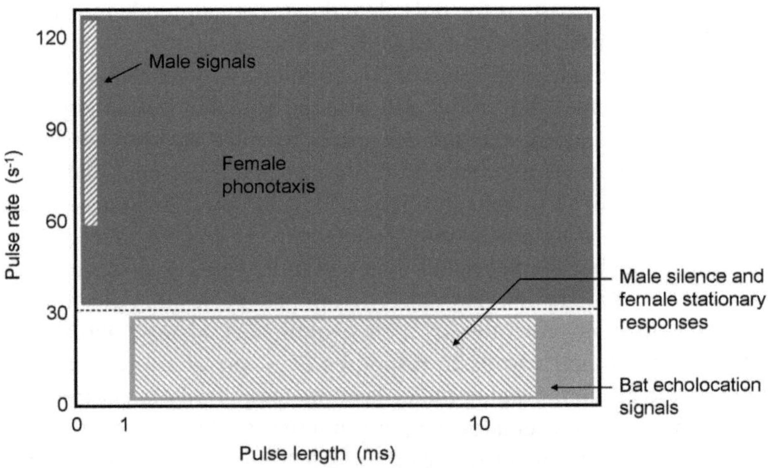

Fig. 6.5 Diagram of pulse pair and pulse rate combinations showing the range of *Achroia grisella* male song and echolocation signals of gleaning bats during their searching phase. Female responses to male song and defensive responses of male and female *A. grisella* to bat echolocation signals while on the substrate are indicated (adapted from Greenfield and Hohendorf 2009). Defensive responses while in flight are given to pulse lengths >1 ms regardless of pulse rate

defensive responses to pulsed ultrasound represent independent traits or different expressions of the same behavioral trait evoked under distinct circumstances? Similarly, do the male silence and female stationary response represent independent or linked traits? Quantitative analysis of inbred lines indicated no evidence of genetic covariance between sexual and defensive behavior (Greenfield and Hohendorf 2009), and in females the QTL that influence these two behaviors are not found on the same chromosomes (Alem 2012). On the other hand, the male silence and the female stationary responses do appear to covary (Greenfield and Hohendorf 2009). That is, they may be sex-specific expressions of the same trait, an inhibition of motor activity in response to stimuli that are normally associated with potential predators.

6.6 Evolution of Acoustic Communication: On the Role of the Sensory Bias Mechanism

The acoustic Lepidoptera offer us a valuable opportunity to probe the evolution of signals in animal communication. Because acoustic signaling used in mating communication in Lepidoptera is so rare while hearing is widespread, the evolution of perception prior to communicative signaling is inferred: One can readily conceive of the occurrence of perception in the absence of communication, and the occurrence of non-communicative acoustic signals in the absence of hearing is equally possible, as in the case of startle or aposematic signals intended for predators. However, communication requires the occurrence of both perception and signaling, with the specific stipulation that these two functions be mutually beneficial to both parties, sender and receiver. Where one function, i.e., hearing, is common while the other, i.e., acoustic signaling, is rare, it is more parsimonious to assume that the rare one evolved later in several clades as opposed to the alternative where the rare one was secondarily lost in the majority of clades.

The above evolutionary sequence implies that acoustic communication in Lepidoptera originated via the 'sensory bias' mechanism (sensu Ryan et al. 1990; Endler and Basolo 1998). That is, male acoustic signals originated inadvertently at some point after hearing evolved. These novel signals would have had a responsive audience—females equipped with tympanal organs and exhibiting certain behavioral reactions to pulsed ultrasound—and thus communication would have appeared without any parallel evolution on the receiver side. It is this absence of receiver evolution that distinguishes the sensory bias mechanism from the various 'co-evolutionary mechanisms', e.g., Fisherian, good genes, chase-away, proposed for signal evolution. The sensory bias mechanism is further supported by the homology of ears within the main acoustic clades, Pyraloidea and Noctuoidea, combined with the multiple, non-homologous structures used for sound production within these clades. This distribution of morphologies is consistent with the proposition that sound production originated on multiple, independent occasions, occurring after hearing had evolved.

Whereas the general notion that hearing preceded sound production in the Lepidoptera may be relatively solid, the details of how acoustic communication evolved remain problematic. Those initial, inadvertent sounds produced by males would have had a female audience, but we can infer that the behavioral responses of these females were negative in terms of movement and orientation: Erratic or evasive flight and dropping to the ground if in the air, and becoming stationary if on the substrate, the basic responses to bat echolocation signals. Thus, we are faced with the challenge of deciphering the evolutionary trajectory from defensive to sexual responses in females. Otherwise stated, how could the males producing these initial, inadvertent sounds have benefited from defensive responses evoked in females such that male sound production was selectively favored? And once favored, how did these inadvertent sounds evolve toward in specialized forms of mating communication?

Various cases reviewed here, particularly those involving moth-bat interactions, provide some insight to the question of how acoustic communication may have emerged from its presumed beginnings. In some arctiid moths, the same structures and sounds are used in defensive and sexual contexts. These male courtship sounds are not identical to bat echolocation signals, but they are of similar high frequency and delivered in short pulses. This overlap suggests that the response evoked in females by the male sounds could resemble that evoked by the bat signals. Unfortunately we have relatively little information on how moths resting on the substrate, which is where courtship in arctiid moths takes place, respond to bat echolocation signals. Whereas the male sounds are known to be necessary for successful courtship in some arctiid species, the specific behavioral responses that females make to these sounds have generally not been directly studied via playback of synthetic signals, which would allow one to disentangle the effects of sound from other male signals. Consequently, we can only suggest that at some point in evolution certain males began to broadcast their defensive sounds when engaged in courtship.

Noctuid and crambid moths are not reported to broadcast sounds intended for bats, but recent findings on several species wherein males produce very low amplitude courtship songs allow comparisons of sexual and defensive responses. In *Spodoptera litura* playback of synthetic male signals to females evoked a stationary response, the freezing of all movement, similar to that observed in response to synthetic bat echolocation signals (Nakano et al. 2010a). Similar behavior also occurred in *Ostrinia furnacalis* (Nakano et al. 2010b). The implication is that females do respond to the male signals as if they were made by predatory bats, and males may benefit from this evoked defensive response because a female who remains motionless may be more easily courted.

The case of *A. grisella* (Pyralidae) is more challenging but must ultimately be analyzed to understand the evolution of acoustic communication. Female *A. grisella* do not simply freeze in response to male song, as they do when exposed to bat echolocation signals, but rather run toward the singing male. Assuming that the sensory bias mechanism accounts for the origin of acoustic communication in moths, we are faced with an inevitable modification of this evolutionary mechanism in which some change occurred in the female subsequent to male signaling. This change would have been essential for *A. grisella* females in their discrimination of male and bat signals. *A. grisella* females distinguish male song from bat echolocation signals primarily on the basis of the slow pulse rate (10–25 pulses·s^{-1}) of echolocation signals of gleaning bats. Were male signals delivered at a high pulse rate at the origin of signaling or did male pulse rate gradually evolve toward faster values alongside a female preference for faster and faster song? Was the hypothetical acceleration of pulse rate in the moth song driven by the need to be distinguished from bat echolocation signals, by sexual selection, or both? And, was the initial male song a crude imitation of bat signaling that elicited a freezing response in females by virtue of its slow pulse rate? Finally, did male pheromones allow females to discriminate males from bats when song originated? Although we cannot answer these questions definitively, it is rather certain that some modification in the female response had occurred: Female *A. grisella* exhibit an abrupt, categorical shift from a defensive to a sexual response as the rate of synthetic ultrasound pulses exceeds 30–40 pulses·s^{-1} (Greig and Greenfield 2004), a feature that would not have been expected in the ancestral state where only defensive behavior occurred.

Is this scenario presented for *A. grisella* song the exception or the rule for the evolution of signaling via sensory bias? It is tempting to propose the latter, as the need to discriminate the signaler from environmental cues will always be present (Greenfield 2002). One approach that could help to resolve questions on signal evolution via the sensory bias mechanism is analysis of genetic architecture, both quantitative and molecular. Genetic architecture has seldom been considered in this context, but information on the genes that influence female responses to male signals and environmental cues could indicate whether these two perceptual traits are controlled pleiotropically by the same locus (loci) (cf. Fuller 2009). Such control is a critical issue, as a 'genetic imprint' in the form of pleiotropy is predicted should the sensory bias mechanism occur in its simplest form: A male signal arises that elicits a female response to environmental cues, and this response remains fixed. However, if some change has occurred in the female response after the origin of the male signal, the genetic loci controlling the two perceptual traits may not be fully identical. Our findings on the genetics of female defensive and sexual responses in *A. grisella* are consistent with the latter situation, male acoustic signals that originated via the sensory bias mechanism and subsequent evolution of the female auditory response to those signals. Here, the genetic analysis supports the general expectation that the female sexual response will not remain unchanged over evolutionary time.

6.7 Conclusion

This chapter has presented the Lepidoptera as an eminently appropriate group in which to probe evolution of acoustic communication, and the sensory bias mechanism in particular. Acoustically communicating Lepidoptera reveal the phylogenetic imprint of the sensory bias mechanism in the evolution of male song, but they also present the dilemma of an improbable transition in the female response. Some resolution of this problem appears possible via further comparative and genetic analyses. It is hoped that evolutionary biologists will avail themselves of the unique opportunities afforded by acoustic Lepidoptera and that further inquiry of signal evolution is undertaken among these very special acoustic species.

Acknowledgments I thank the U.S. National Science Foundation (grants IBN-9407162, DBI-0097223, IBN-9807915, DEB-0131708, IOB-0516634) and the Agence Nationale de la Recherche de France (ANR-07-BLAN-0113-01) for supporting research on acoustic communication in *Achroia grisella*.

References

Alcock J, Bailey WJ (1995) Acoustical communication and the mating system of the Australian whistling moth *Hecatesia exultans* (noctuidae: agaristinae). J Zool 237:337–352

Alcock J, Gwynne DT, Dadour IR (1989) Acoustic signaling, territoriality, and mating in whistling moths, *Hecatesia thyridion* (Agaristidae). J Insect Behav 2:27–37

Alem S (2012) Du jeu des sélections chez une pyrale musicienne: Étude de l'origine, de l'évolution et du maintien sélectif des comportements sexuels. Ph.D. Dissertation. Univ. François Rabelais de Tours, France

Alem S, Greenfield MD (2010) Economics of mate choice at leks: do female waxmoths pay costs for indirect genetic benefits? Behav Ecol 21:615–625

Alem S, Koselj K, Siemers B, Greenfield MD (2011) Bat predation and the evolution of leks in acoustic moths. Behav Ecol Sociobiol 65:2105–2116

Bailey WJ (1978) Resonant wing systems in Australian whistling moth *Hecatesia* (Agarasidae: Lepidoptera). Nature 272:444–446

Bennett AL, Atkinson PR, LaCroix NJS (1991) On communication in the African sugarcane borer, *Eldana saccharina* Walker (Lepidoptera, Pyralidae). J Entomol Soc S Afr 54:243–259

Brandt LSE, Greenfield MD (2004) Condition-dependent traits and the capture of genetic variance in male advertisement song. J Evol Biol 17:821–828

Brandt LSE, Ludvar BCh, Greenfield MD (2005) Co-occurrence of acceptance thresholds and preference functions in female choice: mate discrimination in the lesser wax moth. Ethology 111:609–625

Brunel-Pons O, Alem S, Greenfield MD (2011) The complex auditory scene at leks: balancing anti-predatory behaviour and competitive signalling in an acoustic moth. Anim Behav 81:231–239

Collins RD, Jang Y, Reinhold K, Greenfield MD (1999) Quantitative genetics of ultrasonic advertisement signalling in the lesser waxmoth, *Achroia grisella* (Lepidoptera: Pyralidae). Heredity 83:644–651

Conner WE (1987) Ultrasound—its role in the courtship of the arctiid moth, *Cycnia tenera*. Experientia 43:1029–1031

Conner WE (1999) 'Un chant d'appel amoureux': acoustic communication in moths. J Exp Biol 202:1711–1723

Conner WE, Corcoran AJ (2012) Sound strategies: the 65-million-year-old battle between bats and insects. Annu Rev Entomol 57:21–39

Cremer S, Greenfield MD (1998) Partitioning the components of sexual selection: attractiveness and agonistic behavior in male wax moths, *Achroia grisella* (Lepidoptera: Pyralidae). Ethology 104:1–9

DaCosta MA, Larson P, Donahue JP, Weller SJ (2006) Phylogeny of milkweed tussocks (Arctiidae: Arctiinae: Phaegopterini) and its implications for evolution of ultrasound communication. Ann Entomol Soc Am 99:23–42

Dahm KH, Meyer D, Finn WE, Reinhold V, Röller (1971) Olfactory and auditory mediated sex attraction in Achroia grisella (Fabr). Naturwissenschaften 58:265–266

Davis DR (1998) A world classification of the Harmacloninae, a new subfamily of Tineidae (Lepidoptera: Tineoidea). Smithson Contrib Zool 597:1–57

Dunning DC, Roeder KD (1965) Moth sounds and insect catching behavior of bats. Science 147:173–174

Endler JA, Basolo AL (1998) Sensory ecology, receiver biases and sexual selection. Trends Ecol Evol 13:415–420

Fullard JH (1994) Auditory changes in noctuid moths endemic to a bat-free habitat. J Evol Biol 7:435–445

Fullard JH, Dawson JW, Otero LD, Surlykke A (1997) Bat-deafness in day-flying moths (Lepidoptera: Notodontidae: Dioptinae). J Comp Physiol A 181:477–483

Fuller RC (2009) A test of the critical assumption of the sensory bias model for the evolution of female mating preference using neural networks. Evolution 63:1697–1711

Greenfield MD (1981) Moth sex pheromones: an evolutionary perspective. Fla Entomol 64:4–17 (Proceedings of Insect Behavioral Ecology Symposium, 1980)

Greenfield MD (2002) Signalers and receivers: mechanisms and evolution of arthropod communication. Oxford University Press, Oxford, U.K

Greenfield MD, Baker M (2003) Bat avoidance in non-aerial insects: the silence response of signaling males in an acoustic moth. Ethology 109:427–442

Greenfield MD, Coffelt JA (1983) Reproductive behaviour of the lesser waxmoth, *Achroia grisella* (Pyralidae: Galleriinae): signalling, pair formation, male interactions, and mate guarding. Behaviour 84:287–315

Greenfield MD, Hohendorf H (2009) Independence of sexual and anti-predator perceptual functions in an acoustic moth : implications for the receiver bias mechanism in signal evolution. Ethology 115:1137–1149

Greenfield MD, Medlock C (2007) Temperature coupling as an emergent property: parallel thermal effects on male song and female response do not contribute to species recognition in an acoustic moth. Evolution 61:1590–1599

Greenfield MD, Weber T (2000) Evolution of ultrasonic signalling in wax moths: discrimination of ultrasonic mating calls from bat echolocation signals and the exploitation of an anti-predator receiver bias by sexual advertisement. Ethol Ecol Evol 12:259–279

Grieg EI, Greenfield MD (2004) Sexual selection and predator avoidance in an acoustic moth: discriminating females take fewer risks. Behaviour 141:799–815

Gwynne DT, Edwards ED (1986) Ultrasound production by genital stridulation in *Syntonarcha iriastis* (Lepidoptera: Pyralidae): long-distance signaling by male moths. Zool J Linn Soc 88:363–376

Hauser MD (1996) The evolution of communication. The MIT Press, Cambridge, MA

Heller K-G, Achmann R (1995) Ultrasound communication in the pyralid moth species *Symmoracma minoralis* (Lepidoptera, Pyralidae, Nymphulinae). Entomol Gen 20:1–9

Heller K-G, Krahe R (1994) Sound production and hearing in the pyralid moth *Symmoracma minoralis*. J Exp Biol 187:101–111

Hoy RR, Robert D (1996) Tympanal hearing in insects. Annu Rev Entomol 41:433–450

Jang Y, Greenfield MD (1996) Ultrasonic communication and sexual selection in waxmoths: female choice based on energy and asynchrony of male signals. Anim Behav 51:1095–1106

Jang Y, Greenfield MD (1998) Absolute versus relative measurements of sexual selection:assessing the contributions of ultrasonic signal characters to mate attraction in lesser wax moths, *Achroia grisella* (Lepidoptera: Pyralidae). Evolution 52:1383–1393

Jang Y, Greenfield MD (2000) Quantitative genetics of female choice in an ultrasonic pyralid moth, *Achroia grisella*: variation and evolvability of preference along multiple dimensions of the male advertisement signal. Heredity 84:73–80

Jang Y, Collins RD, Greenfield MD (1997) Variation and repeatability of ultrasonic sexual advertisement signals in *Achroia grisella* (Lepidoptera: Pyralidae). J Insect Behav 10:87–98
Jia F-Y, Greenfield MD, Collins RD (2001) Ultrasonic signal competition among male wax moths. J Insect Behav 14:19–33
Kettlewell HBD (1978) *Acherontia atropos* L., the death's head hawkmoth. Entomol Rec J Var 90:94
Kindl J, Kalinova B, Cervenka M, Jelik M, Valterova I (2011) Male moth songs tempt females to accept mating: the role of acoustic and pheromonal communication in the reproductive behaviour of *Aphomia sociella*. PloS ONE 6(10):e26476
Kitching IJ (2003) Phylogeny of the death's head hawkmoths, *Acherontia* (Laspeyres), and related genera (Lepidoptera: Sphingidae: Sphinginae: Acherontiini). Syst Entomol 28:71–88
Krasnoff SB, Yager DD (1988) Acoustic response to a pheromonal cue in the arctiid moth *Pyrrharctia isabella*. Physiol Entomol 13:433–440
Künike G (1930) Zur biologie der kleinen wachsmotte, *Achroia grisella* (Fabricius). Zeitschrift für angewandte entomologie. 16:304–356
Laboulbène A (1864) Sur l'organe musical de la *Chelonia pudica*. Ann Soc Entomol Fr 4:689–704
Lafaille M, Bimbard G, Greenfield MD (2010) Risk trading in mating behavior: forgoing anti-predator responses reduces the likelihood of missing terminal mating opportunities. Behav Ecol Sociobiol 64:1485–1494
Limousin D (2011) Sélection sexuelle chez un papillon acoustique: étude évolutive du signal mâle et de la préférence femelle. Ph.D. Dissertation. Univ. François Rabelais de Tours, France
Limousin D, Greenfield MD (2009) Evaluation of amplitude in male song: female waxmoths respond to fortissimo notes. J Exp Biol 212:4091–4100
Limousin D, Streiff R, Courtois B, Dupuy V, Alem S, Greenfield MD (2012) Genetic architecture of sexual selection: QTL mapping of male song and female receiver traits in an acoustic moth. PLoS ONE 7(9):e44554. doi:10.1371/journal.pone.0044554
McLachlan R (1877) Letter. Nature 15:254
Minet J (1983) Étude morphologique et phylogénétique des organes tympaniques des Pyraloidea. 1. généralités et homologies. (Lep. Glossata). Ann Soc Entomol Fr 19:175–207
Minet J (1985) Étude morphologique et phylogénétique des organes tympaniques des Pyraloidea. 2. Pyralidae; Crambidae, première partie. (Lepidoptera. Glossata). Ann Soc Entomol Fr 21:69–86
Monge-Najera J, Hernandez F, Gonzalez MI, Soley J, Araya J, Zolla S (1998) Spatial distribution, territoriality and sound production by tropical butterflies (*Hamadryas*, Lepidoptera: Nymphalidae): implications for the industrial melanism debate. Rev Biol Trop 46:297–330
Moritz RFA, Kirchner WH, Crewe RM (1991) Chemical camouflage of the death's head hawkmoth (*Acherontia atropos* L.) in honeybee colonies. Naturwissenschaften 78:179–182
Mutanen M, Wahlberg N, Kaila L (2010) Comprehensive gene and taxon coverage elucidates radiation patterns in moths and butterflies. Proc Roy Soc Lond B 277:2839–2848
Nakano R, Skals N, Takanashi T, Surlykke A, Koike T, Yoshida K, Maruyama H, Tatsuki S, Ishikawa Y (2008) Moths produce extremely quiet ultrasonic courtship songs by rubbing specialized scales. Proc Natl Acad Sci USA 105:11812–11817
Nakano R, Takanashi T, Fujii T, Skals N, Surlykke A, Ishikawa Y (2009a) Moths are not silent, but whisper ultrasonic courtship songs. J Exp Biol 212:4072–4078
Nakano R, Ishikawa Y, Sadahiro T, Skals N, Surlykke A, Takanashi T (2009b) Private ultrasonic whispering in moths. Comm Integ Biol 2(2):123–126
Nakano R, Takanashi T, Skals N, Surlykke A, Ishikawa Y (2010a) To females of a noctuid moth, male courtship songs are nothing more than bat echolocation calls. Biol Lett 6:582–584
Nakano R, Takanashi T, Skals N, Surlykke A, Ishikawa Y (2010b) Ultrasonic courtship songs of male Asian corn borer moths assist copulation attempts by making the females motionless. Physiol Entomol 35:76–81

Nakano R, Takanashi T, Ihara F, Mishiro K, Toyama M, Ishikawa Y (2012a) Ultrasonic courtship song of the yellow peach moth, *Conogethes punctiferalis* (Lepidoptera: Crambidae). Appl Entomol Zool 47:87–93

Nakano R, Ihara F, Mishiro K, Toyama M (2012b) Male courtship ultrasound produced by mesothoracic tymbal organs in the yellow peach moth *Conogethes punctiferalis* (Lepidoptera: Crambidae). Appl Entomol Zool 47:129–135

Reinhold K, Greenfield MD, Jang Y, Broce A (1998) Energetic cost of sexual attractiveness: ultrasonic advertisement in waxmoths. Anim Behav 55:905–913

Rodriguez RL, Greenfield MD (2003) Genetic variance and phenotypic plasticity in a component of female mate choice in an ultrasonic moth. Evolution 57:1304–1313

Rodriguez RL, Greenfield MD (2004) Behavioral context regulates dual function of hearing in ultrasonic moths: bat avoidance and pair formation. Physiol Entomol 29:159–168

Rodriguez RL, Cocroft RB, Schul J, Greenfield MD (2005) Spectral and temporal resolution in the mechanical processing of ultrasonic mating signals in lesser waxmoths. J Exp Biol 208:4159–4165

Rodriguez-Loeches L, Barro A, Perez M, Coro F (2009) Anatomic and acoustic sexual dimorphism in the sound emission system of *Phoenicoprocta capistrata* (Lepidoptera: Arctiidae). Naturwissenschaften 96:531–536

Rowland E, Schaefer PW, Belton P, Gries G (2011) Evidence for short-range sonic communication in lymantriine moths. J Insect Physiol 57:292–299

Ryan MJ, Fox J, Wilczynski W, Rand AS (1990) Sexual selection for sensory exploitation in the frog *Physalaemus pustulosus*. Nature 343:66–67

Sanderford MV (2009) Acoustic courtship in the Arctiidae. In: Conner WE (ed) Tiger moths and woolly bears. Oxford University Press, Oxford, U.K., pp 193–206

Sanderford MV, Conner WE (1995) Acoustic courtship communication in *Syntomeida epilais* Wlk. (Lepidoptera, Arctiidae, Ctenuchinae). J Insect Behav 8:19–31

Sanderford MV, Conner WE (1998) Courtship behavior in *Empyreuma affinis* Roths. (Lepidoptera, Arctiidae, Ctenuchinae): acoustic signals and tympanic organ response. Naturwissenschaften 85:82–87

Scoble MJ (1992) The Lepidoptera: form, function and diversity. Oxford University Press, Oxford, U.K

Shaw KL, Parsons YM, Lesnick SC (2007) QTL analysis of a rapidly evolving speciation phenotype in the Hawaiian cricket *Laupala*. Mol Ecol 16:2879–2892

Simmons RB, Conner WE (1996) Ultrasonic signals in the defense and courtship of *Euchaetes egle* Drury and *E. bolteri* Stretch (Lepidoptera: Arctiidae). J Insect Behav 9:909–919

Skals N, Surlykke A (1999) Sound production by abdominal tymbal organs in two moth species: the green silverline and the scarce silverline (Noctuoidea: Nolidae: Chloephorinae). J Exp Biol 202:2937–2949

Spangler HG (1985) Sound production and communication by the greater wax moth (Lepidoptera, Pyralidae). Ann Entomol Soc Am 78:54–61

Spangler HG (1986) Functional and temporal analysis of sound production in Galleria mellonela (Lepidoptera, Pyralidae). J Comp Physiol A 159:751–756

Spangler HG (1987) Ultrasonic communication in *Corcyra cephalonica* (Stainton) (Lepidoptera, Pyralidae). J Stor Prod Res 23:203–211

Spangler HG, Takessian A (1983) Sound perception by 2 species of wax moths (Lepidoptera, Pyralidae). Ann Entomol Soc Am 76:94–97

Spangler HG, Greenfield MD, Takessian A (1984) Ultrasonic mate calling in the lesser waxmoth. Physiol Entomol 9:87–95

Surlykke A, Gogala M (1986) Stridulation and hearing in the noctuid moth *Thecophora fovea* (Tr). J Comp Physiol A 159:267–273

Surlykke A, Skals N, Rydell J, Svensson M (1998) Sonic hearing in a diurnal geometrid moth, *Archiearis parthenias*, temporally isolated from bats. Naturwissenschaften 85:36–37

Takanashi T, Nakano R, Surlykke A, Tatsuta H, Tabata J, Ishikawa Y, Skals N (2010) Variation in courtship ultrasounds of three Ostrinia moths with different sex pheromones. PloS ONE 5(10):e13144

Tree of Life Web Project (2010) Ditrysia. Version 17 November 2010 (temporary). http://tolweb.org/Ditrysia/11868/2010.11.17 in The Tree of Life Web Project, http://tolweb.org/

Trematerra P, Pavan G (1995) Ultrasound production in the courtship behavior of *Ephestia cautella* (Walk.), *E. kuehniella* Z. and *Plodia interpunctella* (Hb.) (Lepidoptera, Pyralidae). J Stor Prod Res 31:43–48

Weller SJ, Jacobson NL, Conner WE (1999) The evolution of chemical defences and mating systems in tiger moths (Lepidoptera: Arctiidae). Biol J Linn Soc 68:557–578

Yack JE, Fullard JH (2000) Ultrasonic hearing in nocturnal butterflies. Nature 403:265–266

Yager DD (1999) Structure, development, and evolution of insect auditory systems. Microscop Res Tech 47:380–400

Zhou Y, Kuster HK, Pettis JS, Danka RG, Gleason JM, Greenfield MD (2008) Reaction norm variants for male calling song in populations of *Achroia grisella*(Lepidoptera: Pyralidae): towards a resolution of the lek paradox. Evolution 62:1317–1334

Zhou Y, Kelly JK, Greenfield MD (2011) Testing the Fisherian mechanism: examining the genetic correlation between male song and female response in waxmoths. Evol Ecol 25:307–329

Chapter 7
Cicada Acoustic Communication

Paulo J. Fonseca

Abstract Cicadas are iconic insects that use conspicuously loud and often complexly structured stereotyped sound signals for mate attraction. Focusing on acoustic communication, we review the current data to address two major questions: How do males generate specific and intense acoustic signals and how is phonotactic orientation achieved? We first explain the structure of the sound producing apparatus, how the sound is produced and modulated and how the song pattern is generated. We then describe the organisation and the sensitivity of the auditory system. We will highlight the capabilities of the hearing system in frequency and time domains, and deal with the directionality of hearing, which provides the basis for phonotactic orientation. Finally, we focus on behavioural studies and what they have taught us about signal recognition.

This work is dedicated to Franz Huber and Axel Michelsen for teaching me so much....

7.1 Introduction

About 2,500 species of cicadas live in temperate and tropical regions around the world. Among insects they are notorious for their conspicuous loud and complex sound signals, which are stereotyped and species-specific (e.g. Fonseca 1991; but see Sueur and Aubin 2003, Sueur et al. 2007). Their particular temporal and spectral structure depends on the biomechanics of the sound production apparatus, and on the neural networks underlying song pattern generation. The latter determine the timing and bilateral coordination of timbal muscle contractions, i.e. the song pattern (Fonseca et al. 2008).

P. J. Fonseca (✉)
Department of Biologia Animal and Centro de Biologia Ambiental, Faculdade de Ciências, Universidade de Lisboa, Bloco C2, Campo Grande, 1749-016 Lisboa, Portugal
e-mail: pjfonseca@fc.ul.pt

B. Hedwig (ed.), *Insect Hearing and Acoustic Communication*,
Animal Signals and Communication 1, DOI: 10.1007/978-3-642-40462-7_7,

Signals are produced by males for mate attraction, courtship induction, as a distress sound or in the context of male–male interactions (Fonseca 1991). Since mate finding in cicadas is usually mediated by acoustic signals, females must be able to recognise the male signal and to orientate towards the singing male(s). Additionally, a female entering a chorus may need to discriminate among different males in an acoustically noisy environment. The complexity of the courtship behaviour allows females to select a mate, possibly upon multimodal assessment of mechanosensory information, conveyed through airborne sound or substrate vibrations, and other sensory channels such as vision (e.g. Cooley and Marshall 2001). The issue of sexual selection is still poorly addressed in these insects.

Female phonotaxis depends on the sensitivity and directionality of the hearing organs and on the extraction of behaviourally relevant information within the nervous system. Information can be imbedded in amplitude modulations of the male signal and/or in its frequency spectrum. Ultimately, sound signalling and sound reception should co-evolve to allow mate finding. In this context, behavioural studies provide an invaluable tool to reveal subtleties of a species communication system.

Here, we shall address two major questions: First, how do males generate specific and intense acoustic signals and second, how do female cicadas achieve phonotactic orientation? We will explain the structure of the sound producing apparatus, and how the sound is produced and modulated to generate the song pattern. We then will describe the organisation and the sensitivity of the hearing organs. We will highlight the processing capabilities of the auditory pathway in the frequency and time domains, and deal with the directionality of hearing, which provides the basis for phonotactic orientation. Finally, we will focus on behavioural studies and what they have taught us about signal recognition.

7.2 How do Males Generate Specific and Intense Acoustic Signals?

7.2.1 The Structure of the Sound Producing Apparatus

Most male cicadas produce their sound signals through a timbal apparatus (Pringle 1954; Moore and Sawyer 1966; Young and Bennet-Clark 1995; Bennet-Clark 1997; Fonseca and Bennet-Clark 1998). It is located dorso-laterally on both sides in the first abdominal segment (Fig. 7.1) and is generally lacking in females. The central feature of the apparatus is a bi- to multistable convex membrane, the timbal (Fig. 7.1c) with variable thickened sclerotised ribs. In some species, a dorsal bar couples a number of ribs. Small sclerotised patches, the small ribs, may be present between the ribs. Posterior to the ribbed area is the timbal plate where the timbal muscle attaches through a tendon-like structure. The timbal is delimited by a strong rim, the timbal frame. The tensor muscle (Fig. 7.1b, d) inserts and pulls at the anterior region of the timbal frame often

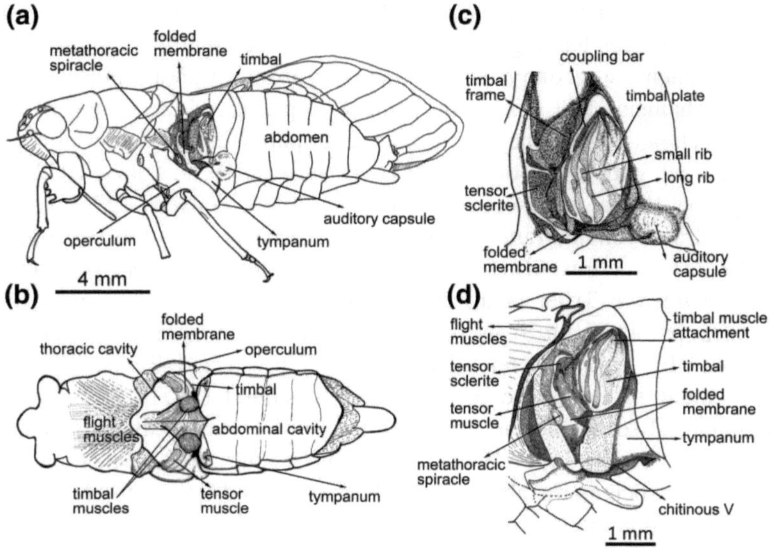

Fig. 7.1 General anatomy of a male cicada. **a** Lateral view with indication of the structures influencing sound production. **b** Schematic longitudinal section revealing the position and extension of the internal air cavities (view from above). **c** External view of the *left* timbal and surrounding structures in *Tettigetta josei*. **d** Longitudinal section depicting an internal view of the *right* timbal apparatus of *T. josei*

differentiated as a tensor sclerite. The frame is surrounded anteriorly and ventrally by a folded membrane which allows for movements of the tensor sclerite as well as dorso–ventral and lateral movements of the abdomen. Internally, a large cavity is formed by fused tracheal air sacs (Fig. 7.1b). These project anteriorly forming a smaller thoracic cavity that backs the timbals and the folded membrane. It connects to the exterior when the prominent metathoracic spiracles open. The posterior cavity fills most of the abdomen. The size of the abdominal cavity and the posture of the abdomen can be varied. Ventrally to the sound producing apparatus and facing frontward lie the tympana of the hearing organs (Fig. 7.1a, b, d) which are described below.

The internal cavities, the structure of the abdominal wall, the tympana, the folded membranes, opercula and timbal covers may all contribute to modify and/or radiate the sound produced by the timbals; their diversity is depicted in Moulds (1990).

7.2.2 Sound Production and Modulation

The primary sound generators are the timbals. When the powerful timbal muscle contracts, the convex timbal is loaded with mechanical energy and eventually collapses inward (Young and Bennet-Clark 1995; Bennet-Clark 1997; Fonseca and

Hennig 1996) allowing a fast energy release. The timbal is driven either in one or in successive steps leading to sequential sudden bending of the long ribs. This inward movement is accompanied by one or a group of sound pulses (IN). Due to these mechanisms and the biomechanical properties of the timbal membrane, the timbal acts as a frequency multiplier. This partly explains how timbal muscle contractions, with a rate of 20–550 Hz (Hagiwara 1955; Young 1972a; Young and Josephson 1983a, b; Josephson and Young 1985; Fonseca 1996), can generate peak call components ranging from about 1 kHz (e.g. *Cystosoma saundersii*, Young 1972a) to ultrasonic frequencies (e.g. *Tettigetta josei*: Fonseca 1991, *Cicadetta iphigenia*: Trilar et al. 2006). Upon relaxation of the timbal muscle, the timbal pops out again to its resting position driven by elastic energy stored in the timbal by resilin (Bennet-Clark 1997; Fonseca and Bennet-Clark 1998). This movement may also be accompanied by sound (OUT). In addition, the timbals load and set in motion the air in the internal cavity that can create cavity resonances (Young 1990; Bennet-Clark and Young 1992, 1998) and/or drive other structures such as the abdomen wall (Fonseca and Popov 1994). As the timbal apparatus is bilaterally organised, activation of both timbal muscles may range from simultaneous to alternating. The sound pulses generated by each timbal muscle contraction compose the basic song element, i.e. the syllable. The syllables can be repeated over time in groups to form echemes, which may in turn delineate longer and more complex song sequences, the phrases (Fig. 7.6a; cf. Fonseca 1991 for terminology).

The convexity and stiffness of the timbal, and thus the sound generated, may be modified by activity of the tensor muscle (Fig. 7.1c, d) (Pringle 1954; Simmons and Young 1978; Hennig et al. 1994; Fonseca and Hennig 1996).

In species where the abdomen appears to act as a Helmholtz resonator, the abdomen cavity and the gap abdomen-opercula can be adjusted to influence the sound quality (Young 1990; Bennet-Clark and Young 1992, 1998; but see Morse and Ingard 1987, for a detailed description of the physics of an Helmholtz resonator and Bennet-Clark 1999, for a general description of resonance models in insects). In species with thick abdomen walls, sound radiation may be primarily via the tympana (Weber et al. 1988; Young 1990); strong amplitude modulations are correlated with vertical movements of the abdomen.

7.2.3 How is the Song Pattern Generated?

Song specificity is determined by the mechanical characteristics of the sound producing structures, and by the coordinated contraction of the muscles affecting the timbals. The large timbal muscle is innervated by a single large timbal motorneuron (Hagiwara and Watanabe 1956; Simmons 1977; Wohlers et al. 1979; Wohlers and Bacon 1980) whereas at least 2–3 motorneurons innervate the tensor muscle (Wohlers et al. 1979; Popov 1981, and functional evidence by Stokes and Josephson 2004). The fast timbal muscle is neurogenic, i.e. each timbal motorneuron action potential causes a twitch contraction. A remarkable exception is *Platypleura capitata,* in which the timbal muscle appears to be myogenic (Pringle 1954).

Little is known about the organisation of the song pattern generator (SPG) in cicadas. Simmons (1977) found a group of interneurons at the metathoracic-abdominal ganglion complex (MAC) that oscillated at twice the frequency of the timbal motorneuron spikes. These were one quarter of a cycle out of phase. Simmons (1977) concluded that several interneurons were involved in generating the song rhythm and that at least some of these should be non-spiking interneurons. In spite of the indirect evidence that these interneurons might be part of the SPG, current injection rarely changed the waves' frequency in the interneuron or in the timbal motorneuron, and thus apparently did not strongly influence the rhythm generator.

Each timbal motorneuron received input from several interneurons. The motorneuron initiating a sound sequence often swapped after pauses in singing and they may be activated at different phases (Fonseca 1996). The timbal motorneurons do not appear to be directly coupled as current injection in one timbal motorneuron did not produce any recognisable effect on the other timbal motorneuron (Simmons 1977; R.M. Hennig and P.J. Fonseca unpublished).

Preliminary work using extra- and intracellular recording and staining investigated the activity of descending and local inter- and motorneurons during singing in the cicada *Tympanistalna gastrica* (R.M. Hennig and P.J. Fonseca unpublished data). A singing pattern similar to the natural calling song was elicited as an after-effect of electrical brain stimulation (for details of this technique introduced by the late A.V. Popov see Fonseca and Popov 1994; for an oscillogram see Fonseca and Bennet-Clark 1998).

As in the flight of orthoperans (Robertson and Pearson 1982, 1984), no evidence was found that the timbal motorneurons are part of the SPG. The activity of the timbal motorneuron (inset in Figs. 7.2a, 7.4c) starts later than most other neurons (Fig. 7.2a). Depolarising current injection, up to 10 nA, applied to the motorneuron never resulted in singing activation. In contrast, in one large local interneuron depolarisation could result in patterned timbal motorneuron activity (inset in Fig. 7.2a, b), i.e. fictive singing. This omega-shaped neuron in MAC, labelled *Singing Interneuron 1 (Si-Int-1*) never generated spikes, and preceded the activity of the timbal motorneuron by about one cycle (Fig. 7.2a, b). Its arborisations in both hemiganglia overlapped with arborisation of the timbal motorneuron (inset in Fig. 7.2a) which may indicate a connection of *Si-Int-1* with the timbal motorneuron. *Si-Int-1* established excitatory connections with *Si-Int-2,* another non-spiking omega-shaped neuron, which closely followed the activity of *Si-Int-1*. Since *Si-Int-1* and *Si-Int-2* establish excitatory connections (data not shown), they might constitute an important core of the SPG. Interestingly, the amplitude of *Si-Int-1*

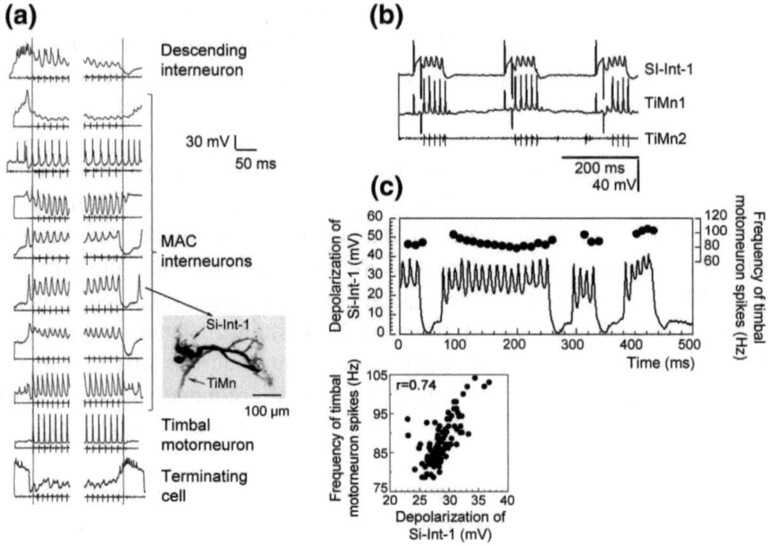

Fig. 7.2 Interneurons active during singing by the cicada *Tympanistalna gastrica*. **a** Comparison of the timing of activation of several interneurons relative to the activity of the timbal motorneuron. Inset is a confocal image of simultaneous Luciffer yellow fills of the local interneuron Singing Interneuron 1 (*Si-Int-1*) and the timbal motorneuron. **b** Details of the activity of the omega-shaped cell *Si-Int-1* in a double intracellular recording with the timbal motorneuron (TiMn1), together with the extracellular recording of the other timbal motorneuron (TiMn2). Singing activity was elicited by depolarising current injection in *Si-Int-1*. Both timbal motorneurons are activated. **c** The representation of the frequency of the timbal motorneuron rhythm superimposed to an intracellular recording of *Si-Int-1*, points to a correlation with the amplitude of depolarisation of *Si-int-1* (graph below)

depolarisation was strongly correlated with the frequency of the timbal motorneuron action potentials (Fig. 7.2c), an observation that deserves further investigation. Differently to Simmons (1977), no evidence was found for continuous oscillation of the SPG activity.

Surprisingly, although both timbal motorneurons are activated almost simultaneously (Fig. 7.2b), EMGs from the timbal muscles reveal a much larger delay (about 2.5 ms, Fonseca 1996), also observed in the sound output. In fact, the IN pulses alternate with a phase ca. 1/6th–1/7th of a period (cf. Fonseca 1996) and singing always initiates with the right timbal (Fonseca and Bennet-Clark 1998). This asymmetry is thus not created at the level of the neuronal oscillators. Instead it must be an attribute of the periphery, probably caused by the asymmetric timbals (Fonseca and Bennet-Clark 1998) that may request different force to buckle inwards.

As in other systems (Hedwig 1996, 2000), it is very likely that singing is elicited by descending activity from command neurons in the brain. Descending neurons whose activation preceded the timbal motorneuron activity by about 2–4 timbal periods could be recorded (Fig. 7.2a).

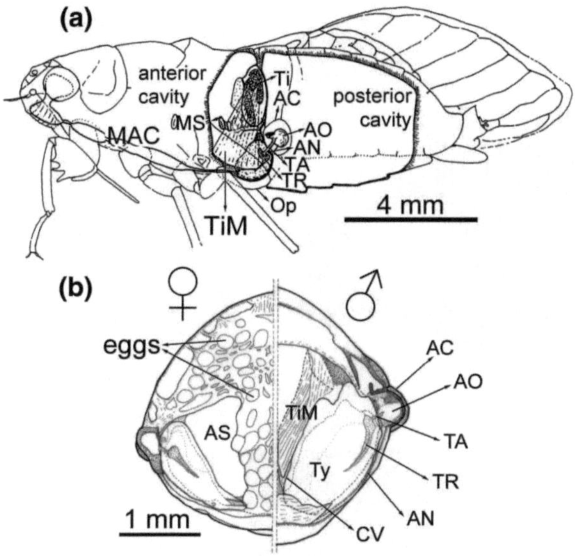

Fig. 7.3 Anatomy of the cicada hearing structures. **a** Diagrammatic lateral view revealing the components of the hearing system and their position relative to the male timbal apparatus. **b** Transverse cut at the level of the auditory organ in a female and a male of the cicada *Tettigetta josei*. AC auditory capsule; AN auditory nerve; AO auditory organ; AS air sac; CV chitinous V; MAC metathoracic-abdominal ganglion complex; MS metathoracic spiracle; Op operculum; TA tympanic apodeme; Ti timbal; TiM timbal muscle; TR tympanic ridge; Ty tympanum

7.3 How is the Auditory System Organised?

7.3.1 Structure of the Hearing Organ

Cicadas have a highly specialised auditory system (Fig. 7.3) with a basic structure similar across species (Vogel 1923; Myers 1928; Pringle 1954; Michel 1975; Young and Hill 1977; Doolan and Young 1981; Fonseca 1993, 1994; Fonseca and Popov 1997). The ears are situated latero-ventrally in the first abdominal segment (Fig. 7.3a, b). The delicate tympana, which are much larger in males than in females, may in parts be under 1 μm thick. They are backed and acoustically coupled by tracheal air cavity(ies) (Fig. 7.3a) that open through the metathoracic spiracles (Figs. 7.1d, 7.3a). The folded membranes integrate the anterior cavity wall which, in males, backs the timbals (Fig. 7.1b, d). The male abdomen can be very thin and translucent. Thus, sound may reach the internal surfaces of the tympana via the contralateral tympanum, the metathoracic spiracles and the folded membranes and, in males, also through the timbals and the thin abdominal wall (Fig. 7.1d). The tension of the tympanum can be varied by the action of detensor tympani muscles (Pringle 1954), which can modulate hearing sensitivity (Hennig

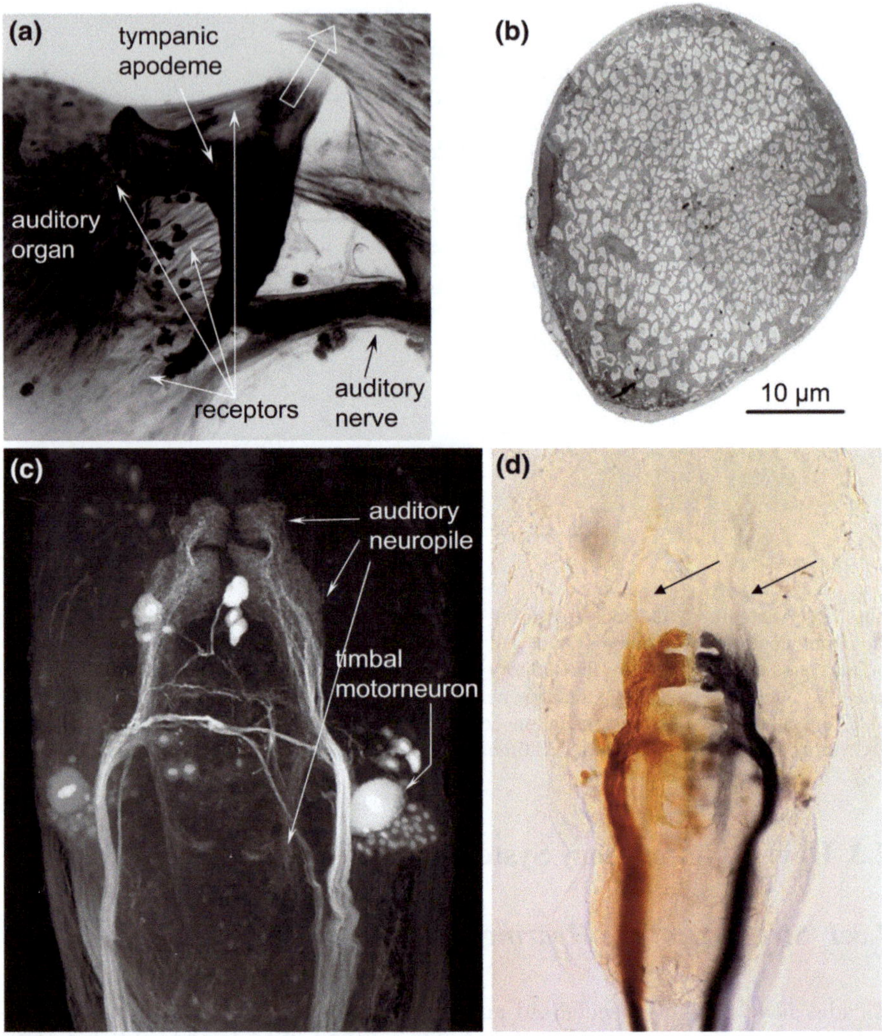

Fig. 7.4 Details of elements in the auditory pathway of the cicada *Tettigetta josei*. **a** Two-photon confocal image showing the distribution of auditory receptor cell bundles and their connection to the complex-shaped crescent-like tip of the auditory apodeme. The axons of the receptors coalesce in several branches that integrate the auditory nerve. The wide arrow indicates the direction of the tympanic apodeme towards the tympanum. **b** Electron microscope view of a transverse cut of the auditory nerve just exiting the auditory organ. Each profile corresponds to one axon of one auditory receptor cell. Many axons exhibit sub-micrometre profiles. **c** Two-photon confocal image of the auditory neuropile revealed by backfilling of both auditory nerves with Lucifer Yellow fluorescent dye. The auditory receptors project into a complex auditory neuropile that spans several segmental areas in the fused metathoracic-abdominal ganglion complex. The two large timbal motorneurons, whose axons run in this, and in many other cicada species, in the auditory nerve, cross at the mid line. **d** In other species, such as in *Cicada barbara*, the timbal motorneuron runs in an independent nerve. Here the auditory neuropile is revealed by backfilling one auditory nerve with nickel chloride (*dark blue*) and the other with cobalt chloride (*dark orange*). Some axons appear not to end at the neuropile (see *arrows*)

et al. 1994). The tympanic ridge, a sclerite that sits on the thin tympanic membrane, connects the tympanum to the auditory organ, which is protected within the auditory capsule, through a lever, the tympanic apodeme (Fig. 7.3b). About 500–2,200 auditory receptors (Fig. 7.4b) attach to the apodeme tip (Fig. 7.4a) and form the onion-shaped auditory organ (Figs. 7.3b, 7.4a) (Wohlers et al. 1979; Doolan and Young 1981; Fonseca 1994). The receptors' axons coalesce in an auditory nerve that joins the MAC (Figs. 7.3b, 7.4a) and project into a complex auditory neuropile (Fig. 7.4c, d) (Wohlers et al. 1979; Fonseca 1994). The auditory nerve may also contain other sensory fibres, and the axon of the timbal motorneuron (Fig. 7.4c).

7.3.2 The Sensitivity of the Auditory System

Only anecdotal data are available from single auditory afferent recordings (e.g. Münch 1999) as axon diameters are in the range of 1 micron (Fig. 7.4b), Sensitivity of the hearing organ has been analysed with recordings of the whole auditory nerve or auditory interneurons. Nerve recordings in different species (Popov 1981; Popov et al. 1985; Huber et al. 1990; Fonseca 1994; Daws and Hennig 1996; Fonseca et al. 2000; Fonseca and Cooley unpublished) revealed threshold curves with sensitivity of 25–40 dB SPL (i.e. re. 20 μPa) to frequencies ranging 2–6 kHz (but see Young and Hill 1977). The best frequency range is the same across many species, even in cases where the songs' spectra are almost devoid of energy at this range. This puzzling mismatch (Popov 1990; Fonseca 1994) is an artefact at least in some species. Intracellular recordings in *Tettigetta josei* (Fonseca et al. 2000) (Fig. 7.5) and *Cicada barbara* (Fonseca 1994), revealed auditory interneurons sensitive to frequencies higher than expected from auditory nerve recordings. This unsuspected interneuron sensitivity may be attributed to auditory afferents with very thin axons (Fig. 7.4b) (Fonseca et al. 2000), which are not properly represented in whole nerve recordings. The ubiquitous strong auditory sensitivity within the low frequency range suggests the activation of a large

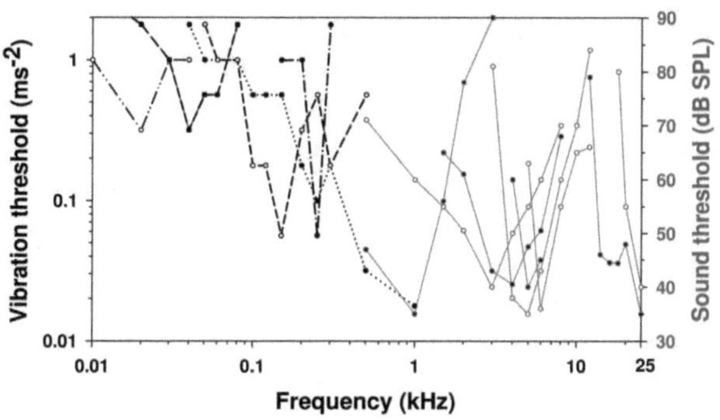

Fig. 7.5 Threshold curves of interneurons of the cicada *Tettigetta josei* sensitive to vibration (*black interrupted* and *thicker lines*) and to sound (*grey solid thinner lines*). This cicada shows interneurons capable of analysing vibrations and sound encompassing a large spectral range

number of afferents, and may be related to selection pressure to detect predators rather than conspecifics (e.g. Popov 1990; Mason 1991).

7.3.3 Processing in the Frequency Domain

Songs vary widely across cicada species and include wide band and pure-tone spectra, ranging from few hundred Hz to ultrasound (e.g. Young 1972b; Popov 1975; Popov et al. 1985; Fonseca 1991, 1994; Gogala 1995; Sueur and Aubin 2003; Sueur et al. 2004; Moulds 2005; Trilar et al. 2006). Songs can show pronounced frequency modulations, especially in tropical species (e.g. Gogala 1995). Based on the very large number of auditory afferents and at least 15 auditory ascending interneurons (Huber et al. 1980; Fonseca et al. 2000; Fonseca and Correia 2007) cicadas may be able to process details in their songs, both in frequency and in time domains.

Tympanic membrane vibrations (Fonseca 1993; Fonseca and Popov 1997; Sueur et al. 2006, 2008, 2010) are transmitted to the onion-shaped auditory organ by the flattened rod-like tympanic apodeme in a way that may contribute to frequency analysis in the auditory pathway. Some underlying mechanisms revealed in *T. josei* point to (1) the complex vibrations observed at the tympanic ridge, (2) the shape of the apodeme tip and (3) the distribution and orientation of the afferent neurons' attachment to the apodeme (Fig. 7.4a) (Michelsen and Fonseca, in preparation). The apodeme originates from, and forms an angle with, the tympanic ridge which vibrates in a complex mode. Depending on frequency it may rock, move up and down or back and forth, movements that are communicated to the apodeme. At the edge of the tympanum, the apodeme barely moves at most frequencies and appears like a lever anchored at an intermediate point, i.e. it inverts and transmits the movement to the crescent-like apodeme tip, which has the freedom to move in three dimensions.

Apodeme geometry may be simpler in other species. Bundles of receptor cells oriented in all three space dimensions attach just before the tip and along the crescent-like structure (Fig. 7.4a). Such special arrangement, also found in the bladder cicada *Cystosoma saundersii* (Doolan and Young 1981), allows the sensory neurons to be maximally activated according to their directions of attachment. Since the orientation of the apodeme movement is frequency dependent, frequency discrimination could be due to the activation of differently oriented receptors. In addition, intrinsic cellular mechanisms were proposed (Fonseca and Correia 2007) and demonstrated in other insects (Göpfert and Robert 2001; Göpfert et al. 2005; Kernan 2007) and in vertebrates (Fettiplace 1987; Dallos 1992; Kennedy et al. 2005). If present in cicadas, such mechanisms could enhance frequency discrimination.

Interneurons seem to represent mechanosensory information from very low frequency substrate-born vibrations to high frequency air-born signals (cf. Fig. 7.5). In *Tettigetta josei,* a set of ascending interneurons with high Q_{10dB} values is tuned to different frequencies in the range 1–25 kHz (Fig. 7.5) (Fonseca et al. 2000). In addition to air-born sounds, cicadas also detect substrate vibrations mostly via subgenual organs since afferent activity is mainly found in the leg's nerve. Interneurons tuned to different frequencies within 20–1,000 Hz with sensitivities ranging from 0.02 to 0.3 ms^{-2} (Fig. 7.5) have been described (Fonseca and Santos 2001) some of which were activated by vibrations induced in a plant during a cicada landing or takeoff. In small-sized cicadas, like *T. josei,* males move around with short flights intercalated by short calling sequences ("sing-fly" behaviour) and females wait deeply within the vegetation until a male lands and sings close by and, only then signal e.g. by wing flicking (e.g. Gwynne1987; Lane 1995; Marshall and Cooley 2000, 2001; Cooley 2001). Therefore substrate vibrations may also be important in intraspecific communication (Fonseca and Santos 2001; Stölting et al. 2002) and may be invaluable to detect approaching predators (Kühne 1982; Kühne et al. 1984; Hill 2001).

7.3.4 Processing in the Time Domain

Cicadas' acoustical signals show great variability in temporal patterns among species (e.g. Popov 1975; Popov et al. 1985; Fonseca 1991; Moulds 1990, 2005; Gogala 1995; Sueur et al. 2004; Gogala and Trilar 2004; Gogala et al. 2005; Trilar et al. 2006). Mate finding is usually mediated by the calling song but final acceptance of a male by the female depends on subtleties of the courtship behaviour (Cooley 1999; Cooley and Marshall 2001), including courtship song. This makes sound an important pre-mating mechanism for species isolation and raises the question to what degree cicadas can process the time pattern of conspecific songs.

Different methods have been used to estimate auditory temporal resolution in animals (Michelsen 1985; Tougaard 1998), but only a few studies have approached the question in cicadas (Huber et al. 1990; Fonseca 1994).

Using the calling song and natural sounds evidence for time resolution in the auditory pathway was obtained in *T. josei* (Alves and Fonseca, unpublished). This species produces calling song phrases with two distinctive parts (Fig. 7.6a); a succession of echemes (part 1) which ends with a more continuous buzz (part 2) (details in Fonseca 1991). This pattern creates sound pulse periods of 3–5 ms within the echemes, echeme periods around 30 ms and periods of 10–15 ms between consecutive loud IN pulses during the buzz (cf. Fig. 7.6a). The representation of these calling song features was investigated at the level of summed auditory nerve activity and auditory interneuron responses. Auditory nerve activity represented any sound pulses down to at least periods of 1 ms provided the first pulse was quieter than the second (Fig. 7.6b, OUT1-IN2), the normal condition in echemes. If the first pulse was considerably louder, as in the last part of an echeme, a weak nerve response occurred for pulse periods of about 4 ms, augmenting with increasing intervals (Fig. 7.6b, IN2-OUT2). Within auditory interneurons, the best response occurred in ascending neurons very sensitive to the calling song (Fig. 7.7a). At least one neuron represented the loud IN pulses of the buzz down to a period of 6 ms (note that the IN pulses in buzzes are always loud). Manipulation of the silent interval between the quiet (OUT1) and the loud

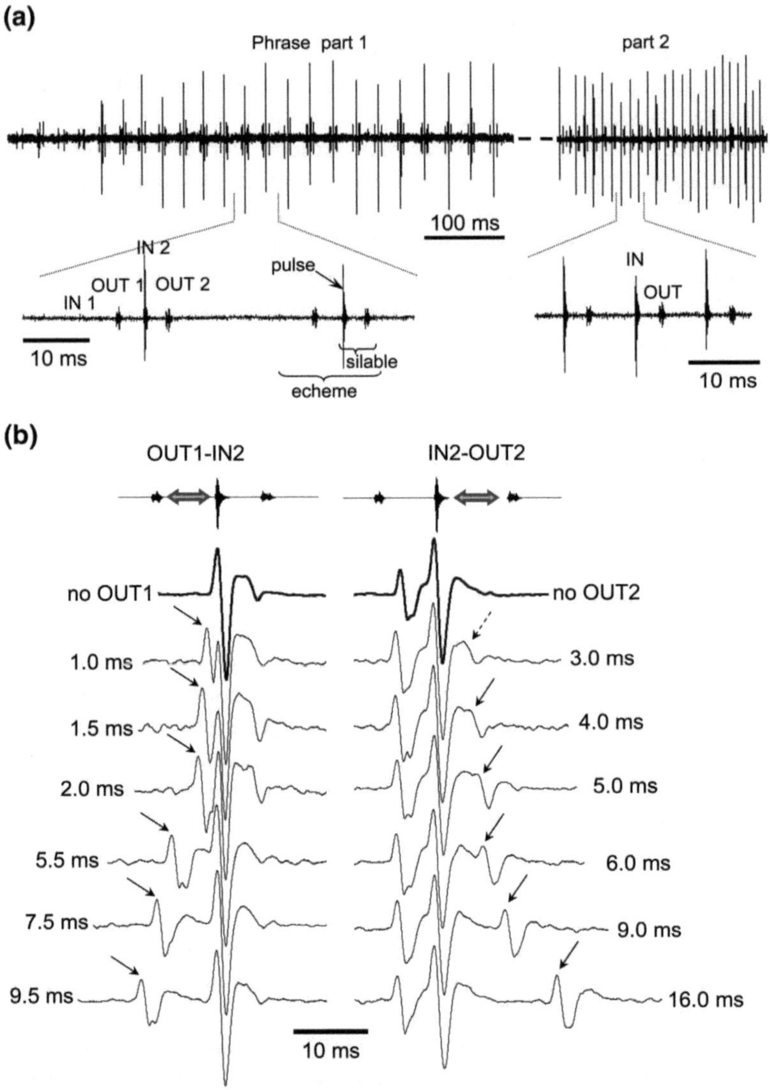

Fig. 7.6 Calling song and temporal resolution of the auditory pathway of the cicada *Tettigetta josei* exhibited by the summed activity of the auditory receptors. **a** The calling song consists in a sequence of temporally complex phrases with two distinct parts. Echemes in part 1 exhibit a characteristic amplitude modulation with the IN sound pulse increasing strongly from the first syllable, where it can be barely noticeable, to the second syllable, where the IN is the loudest pulse. The pattern is different in the second part, where the IN pulses are always loud. This amplitude modulation pattern generates different temporal cues. **b** Nerve recordings represent any timing between two sound pulses provided the first is quitter, a characteristic of the first syllable in part 1 echemes. In the second syllable, when the second sound pulse has a lower amplitude, it appears in the averaged activity of the auditory nerve (average of 16 echemes, $t = 28$ °C) only for periods above 3–4 ms

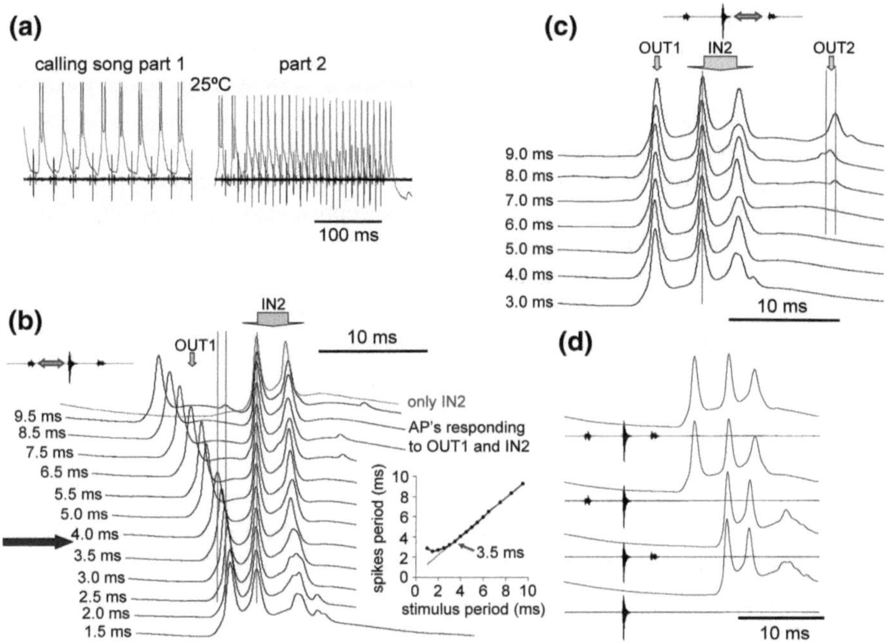

Fig. 7.7 Representation of the temporal characteristics of the song by an auditory ascending interneuron type sensitive to the calling song in the cicada *Tettigetta josei* (averages of 16 stimuli). **a** Song and interneuron activity at 25 °C. The timing of the echemes is well represented in the two parts of a phrase. **b** Manipulation of the period in the first audible two pulses of a part 1 echeme, i.e. Out1-In2, shows that this period is faithfully represented in the spiking activity at least down to 3.5–4 ms (graph inset when $x \approx y$), In contrast, the timing at the end of an echeme is only represented for much longer periods (>7–8 ms) **(c)**. **d** Selectively removing the softer (OUT) pulses of a part 1 echeme unequivocally shows that this pulse is correctly represented in the first syllable of the echeme, but not in the second syllable

(IN2) pulses characteristic of the echemes revealed that the same cell represented the onset of the two pulses down to at least 4 ms (Fig. 7.7b). Similar temporal resolution down to 1–4 ms occurs in other insects (e.g. von Helversen 1979; Ronacher and Stumpner 1988; Tougaard 1996; Prinz and Ronacher 2002; Franz and Ronacher 2002). In contrast, if the quiet pulse followed the loud one, the neuron only responded with some irregular activity if periods were longer than 7–9 ms (Fig. 7.7c). Confirmation that the responses to the first (quieter OUT1) and the second (loud IN2) pulses were actually caused by the onset of the sounds, was further obtained by suppressing the quieter first (OUT1), the quieter last (OUT2) or both of them keeping only the loud (IN2) pulse (Fig. 7.7d). This unequivocally ruled out that responses to the first two pulses in an echeme (i.e. OUT1–IN2) might be artefacts created by sustained spiking activity due to suprathreshold stimulation. It demonstrated that the first two pulses in an echeme, OUT1–IN2, but not the last OUT2 were represented in the neuron's spiking activity. A conceivable physiological reason for the representation of two pulses with the second being louder than the first was advanced by Münch (1999).

These results were obtained with recordings at 25–28 °C and the performance reduced with lower temperatures. The temporal resolution measured in the experiments, however, corresponds to temporal characteristics of songs produced at higher temperatures (Fonseca and Allen-Revez 2002a), indicating that all time periods within the calling song of *T. josei* are represented by ascending neurons and forwarded to the brain. However, more electrophysiological and behavioural work is needed to reveal the capabilities of ascending interneurons and especially of auditory brain neurons to deal with temporal features of the songs. The latter question may also be approachable by behavioural experiments.

7.3.5 Directional Hearing

Localisation of a sound source depends primarily on the ability of the peripheral auditory organs to be differently activated according to sound direction, and/or on the ability to process the time difference of the sound arrival at both hearing organs (reviews by Michelsen 1998 and Michelsen and Larsen 2008).

In cicadas, directional hearing was studied in few species by measuring tympanic membrane vibrations (Fonseca 1993; Fonseca and Popov 1997; Fonseca and Hennig 2004) or auditory nerve activity (Young and Hill 1977; Fonseca 1994; Daws and Hennig 1996; Fonseca and Hennig 2004). Significant directionality occurred both at low frequencies and around the peak of the calling song spectrum, with the exception of *C. saundersii* males, where no directionality was found (Young and Hill 1977). Experiments with selective and reversible blocking of putative sound inputs to the auditory system (Fonseca 1993; Fonseca and Popov 1997) indicated that in males the sound generating timbal acted as an important input responsible for the directionality at the spectral peak of the song, a frequency that corresponds to the natural resonance of the timbal (e.g. Fonseca 1993; Fonseca and Hennig 2004). The input through the contralateral tympanum caused high directionality at middle range frequencies (3–8 kHz) in males of *T. gastrica* and *T. josei* but not in *Cicada barbara* males. Instead, in *C. barbara* males, the hollow and thin abdomen was responsible for a directionality response at low frequencies (1–2 kHz) (Fonseca and Popov 1997). By contrast, the most important acoustic inputs conditioning hearing directionality in females were, in addition to the contralateral tympanum, the metathoracic spiracles. Usually, the auditory directionality of females encompasses a large frequency range starting at low frequencies and includes the loud frequency components of the male song. Interestingly, the contralateral tympanum not only acts as an important sound input, but also determines the correct phase lag between the bilateral sound inputs to create pronounced directionality (Figs. 7.6, 7.7 in Fonseca 1993; see Löhe and Kleindienst 1994; Michelsen and Löhe 1995 for a similar observation involving the central membrane in crickets).

In two cicada species, *T. josei* and *T. gastrica*, the directional differences of tympanic vibrations are encoded by the activity of the auditory afferents. Directionality with frontal sound stimulation at ±30º could differ by more than 5 dB (Fonseca and Hennig 2004). In addition, ascending interneurons in *T, josei* showed threshold differences up to 15 dB with frontal stimulation at ±45°, both at lower frequency ranges (3–6 kHz) and around the calling song spectral peak of 16 kHz (Fonseca, unpublished).

7.3.6 What Behavioural Studies have Taught us About Signal Recognition in Cicadas?

The ultimate answer to questions regarding the capacity of an animal to recognise and orient towards sound signals comes from behavioural experiments. Successful experiments rely on subjects receptive and prone to react to the stimulus, what may depend on their physiological state, e.g. circadian influences (Daws et al. 1997) or the receptivity of the female (Cooley and Marshall 2001). This imposes considerable difficulties if the insects cannot be bred in the laboratory and for cicada probably is a cause for the scarcity of behavioural data, compared to the extensive work involving other acoustically communicating insects. For example, receptive female cicadas, or isolated males, fly towards a singing male or a chorus. However, flight phonotaxis experiments have been difficult to conduct in captivity.

In some cicada species, a stereotyped and well-timed short wing flicking signal is produced by receptive females upon listening to the male song (e.g. Cooley 1999; Cooley and Marshall 2001). Such duets, with males producing loud sounds and females responding with wing flicks, should allow to evaluate the relevant parameters for species-specific song recognition. For instance, in many New Zealand cicadas, the songs were found to possess an introductory section and a cueing section, which is responsible for releasing the wing flick response (Marshall, Hill and Cooley personal communication; cf. Fig. 7.1 in Marshall et al. 2008). Therefore, taxa like the genus *Kikihia* and other Cicadettini may be interesting groups to analyse auditory signal and song recognition. Based on the female wing flick response, Marshall and Cooley (2000) demonstrated that females of the 13-year periodical species *Magicicada tredecim* and *M. neotredecim* responded selectively to the dominant frequency of the species-specific male calling songs; both species have nearly pure-tone calls that lack temporal patterns. The dominant frequency of the male song, but not the temporal pattern, was also found to be key to elicit flight phonotaxis in female *C. saundersii* (Doolan and Young 1989); Daws et al. (1997) argued that this frequency-dependent phonotaxis did not necessarily result from fine frequency selectivity of the females, but rather appeared to be based on the overall level of excitation of the auditory system. In fact, an increase in the amplitude of a stimulus outside the best hearing range compensated the reduction in overall auditory excitation and re-established the level of female phonotaxis. Female courtship responses, however, were only elicited if the temporal parameters of the natural song were present in the stimuli, even when synthesised with different carrier frequencies. Thus, the carrier frequency of calling songs appears to be more important in long range communication, to attract flying females, while details of the temporal structure may be essential for short range courtship interactions (Doolan and Young 1989). In this way, the cicada communication system would circumvent the constraint of random amplitude modulation that inevitably affects distant sound propagation, especially within vegetation (e.g. Richards and Wiley 1980). Behavioural data from *Tibicina haematodes*, which aggregate to form choruses, are in line with these results. Males responded to conspecific as well as to allospecific calling songs with overlapping song spectra but distinct temporal pattern. However, they did not react to playbacks of heterospecific songs with disjunct frequency spectra (Sueur and Aubin 2002). In contrast, an apparent absence of frequency selectivity was found in *Cyclochila australasiae* (Daws et al. 1997).

Evidence for frequency analysis and evaluation of temporal pattern was obtained in males of *C. barbara*, which also aggregate in choruses (Fonseca and Allen-Revez 2002b). The readiness of males to sing once another male initiated singing can be used in behavioural experiments. Males responded to the conspecific song as well as to a continuous pure-tone of 6 kHz, the song's spectral peak component. However, in spite of the species' broadband calling song spectrum, and in spite of the maximal peripheral excitation at 3–4 kHz, the males' stereotyped response decreased significantly when tested with pure-tone stimuli at 3 or 4 kHz. This revealed at least some ability for frequency discrimination among 6, 4 and 3 kHz, compatible with the frequency selectivity found at the level of auditory interneurons in *T. josei* (Fonseca et al. 2000). At 90 dB SPL playback intensities, no significant differences in the response occurred between 6 and 9 kHz. This is in line with the finding that this cicada readily reacts to loud songs of another sympatric species (*Tibicina garricola*), also with a continuous song but a spectral maximum at about 9 kHz, a frequency well represented in the *C. barbara* song. The response at 9 kHz deteriorated if the playback amplitude was lowered by 20 dB. This might be attributed to a considerable lower peripheral excitation at 9 kHz when compared to the excitation at 6 kHz (Fonseca and Allen-Revez 2002b), as argued for *C. saundersii* by Daws et al. (1997); this should also effectively prevent a response to singing males of *T. garricola* singing at a distance in natural conditions.

When pauses were introduced in the song of *C. barbara,* the males' responsiveness was maintained if the pauses were shorter than 30 ms, irrespective of the sound duration. The response decreased steeply when pauses exceeded 30 ms (cf. Fonseca and Allen-Revez 2002b). Only this temporal discrimination prevents a response to the calling song of a sympatric and synchronic sister species (*Cicada orni*), which has a largely overlapping frequency spectrum but is composed by a succession of echemes separated by silent intervals longer than 40 ms (Fonseca 1991). Remarkably, *C. orni* males stop responding to song models with silent intervals shorter than 40 ms (Simões and Quartau 2006), suggesting that the duration of the silent intervals is paramount for species discrimination. Studying the responses of females of these two species to the same playback signals should confirm the importance of the pause length for segregation of these species, but so far has not been possible.

Evidence from the few behavioural studies suggests that cicada species may exhibit different capabilities to extract information from the songs both in time and frequency domains. During evolution, these abilities may have been shaped by male–male competition, by the need to detect acoustic cues of individuals within crowded and noisy choruses (Cooley and Marshall 2001) and by the requirement to recognise species-specific song features in noisy habitats with many sympatric species.

Acknowledgments I thank Berthold Hedwig, John Cooley, Sofia Seabra, Maria Clara Amorim and Axel Michelsen for their comments that improved the manuscript, and José Feijó for the confocal microscope images.

References

Bennet-Clark HC (1997) Tymbal mechanics and the control of song frequency in the cicada *Cyclochila australasiae*. J Exp Biol 200:1681–1694

Bennet-Clark HC (1999) Resonators in insect sound production : how insects produce loud pure-tone songs. J Exp Biol 202:3347–3357

Bennet-Clark HC, Young D (1992) A model of the mechanism of sound production in cicadas. J Exp Biol 173:123–153

Bennet-Clark HC, Young D (1998) Sound radiation by the bladder cicada *Cystosoma saundersii*. J Exp Biol 201:701–715

Cooley JR (1999) Sexual behavior in North American cicadas of the genera *Magicicada* and *Okanagana*. PhD dissertation, University of Michigan, Ann Arbor

Cooley JR (2001) Long-range acoustical signals, phonotaxis, and risk in the sexual pair-forming behaviors of *Okanagana canadensis* and *O. rimosa* (Hemiptera*: Cicadidae*). Ann Ent Soc Am 94:755–760

Cooley JR, Marshall DC (2001) Sexual signaling in periodical cicadas, *Magicicada spp*. Behaviour 138:827–855

Dallos P (1992) The active cochlea. J Neurosci 12:4575–4585

Daws AG, Hennig RM (1995/1996). Tuning of the peripheral auditory system of the cicada, *Cyclochila australasiae*. Zoology 99:175–188

Daws AG, Hennig RM, Young D (1997) Phonotaxis in the cicadas *Cystosoma saundersii* and *Cyclochila australasiae*. Bioacoustics 7:173–188

Doolan JM, Young D (1981) The organization of the auditory organ of the bladder cicada *Cystosoma saundersii*. Phil Trans R soc Lond 291(1055):525–540

Doolan JM, Young D (1989) Relative importance of song parameters during flight phonotaxis and courtship in the bladder cicada *Cystosoma saundersii*. J Exp Biol 141:113–131

Fettiplace R (1987) Electrical tuning of hair cells in the inner ear. Trends Neurosci 10:421–425

Fonseca PJ (1991) Characteristics of the acoustic signals in nine species of cicadas (Homoptera, *Cicadidae*). Bioacoustics 3:173–182

Fonseca PJ (1993) Directional hearing of a cicada: biophysical aspects. J Comp Physiol A 172:767–774

Fonseca PJ (1994) Acoustic communication in cicadas (Homoptera, *Cicadoidea*): sound production and sound reception. PhD dissertation, Universidade de Lisboa, Lisboa

Fonseca PJ (1996) Sound production in cicadas: timbal muscle activity during calling song and protest song. Bioacoustics 7:13–31

Fonseca PJ, Allen Revez M (2002a) Temperature dependence of cicada songs (Homoptera, *Cicadoidea*). J Comp Physiol A 187:971–976

Fonseca PJ, Allen Revez M (2002b) Song discrimination by male cicadas *Cicada barbara lusitanica* (Homoptera, *Cicadidae*). J Exp Biol 205:1285–1292
Fonseca PJ, Bennet-Clark HC (1998) Sound radiation in a cicada: the role of tymbal asymmetry in the production of complex songs. J Exp Biol 201:717–730
Fonseca PF, Correia T (2007) Effects of temperature on tuning of the auditory pathway in the cicada *Tettigetta josei* (Hemiptera, *Tibicinidae*). J Exp Biol 210:1834–1845
Fonseca PJ, Hennig RM (1996) Phasic action of the tensor muscle modulates the calling song in cicadas. J Exp Biol 199:1535–1544
Fonseca PJ, Hennig RM (2004) Directional characteristics of the auditory system of cicadas: is the sound producing tymbal an integral part of directional hearing? Physiol Entomol 29:1–9
Fonseca PJ, Popov AV (1994) Sound radiation in a cicada: the role of different structures. J Comp Physiol A 175:349–361
Fonseca PJ, Popov AV (1997) Directionality of the tympanal vibrations in a cicada: a biophysical analysis. J Comp Physiol 180:417–427
Fonseca PJ, Santos T (2001) Sensitivity of substrate vibrations in the cicada *Tettigetta josei* (Homoptera, *Cicadoidea*). In: Elsner N, Kreutzberg GW (eds) Proceedings 28th gottingen neurobiology conference, vol 2. G Thieme Verlag, Stuttgart, p 374
Fonseca PJ, Münch D, Hennig RM (2000) How cicadas interpret acoustic signals. Nature 405:297–298
Fonseca PJ, Serrão EA, Pina-Martins F, Silva P, Mira S, Quartau JA, Paulo OS, Cancela L (2008) The evolution of cicada songs contrasted with the relationships inferred from mitochondrial DNA (Insecta, Hemiptera). Bioacoustics 18:17–34
Franz A, Ronacher B (2002) Temperature dependence of temporal resolution in an insect nervous system. J Comp Physiol A 188:261–271
Gogala M (1995) Songs of four cicada species from Thailand. Bioacoustics 6:101–116
Gogala M, Trilar T (2004) Biodiversity of cicadas in Malaysia—bioacoustic approach. Serangga (Bangi) 9(1/2):63–81
Gogala M, Trilar T, Krpač V (2005) Fauna of singing cicadas (Auchenorrhyncha: *Cicadoidea*) of Macedonia: a bioacoustic survey. Acta Entomol Slov 13(2):103–126
Göpfert MC, Robert D (2001) Active auditory mechanics in mosquitoes. Proc R Soc Lond B 268:333–339
Göpfert MC, Humphris ADL, Albert JT, Robert D, Hendrich O (2005) Power gain exhibited by motile mechanosensory neurons in *Drosophila* ears. PNAS 102:325–330
Gwynne DT (1987) Sex-biased predation and the risky mate-locating behavior of male tick-tock cicadas (Homoptera: *Cicadidae*). Anim Behav 35:571–576
Hagiwara S (1955) Neuro-muscular mechanism of sound production in the cicada. Physiol Comp Oecol 4:142–153
Hagiwara S, Watanabe A (1956) Discharges in motoneurons of cicada. J Cell Comp Physiol 47:415–428
Hedwig B (1996) A descending brain neuron elicits stridulation in the cricket *Gryllus bimaculatus* (de Geer). Naturwissenschaften 83:428–429
Hedwig B (2000) Control of cricket stridulation by a command neuron: efficacy depends on behavioural state. J Neurophysiol 83:712–722
Hennig RM, Weber T, Huber F, Kleindienst H-U, Moore TE, Popov AV (1994) Auditory threshold change in singing cicadas. J Exp Biol 187:45–55
Hill PSM (2001) Vibration and animal communication: a review. Amer Zool 41:1135–1142
Huber F, Wohlers D, Moore TE (1980) Auditory nerve and interneurone responses to natural sounds in several species of cicadas. Physiol Entomol 5:25–45
Huber F, Kleindienst H-U, Moore TE, Schildberger K, Weber T (1990) Acoustic communication in periodical cicadas: neuronal responses to songs of sympatric species. In: Gribakin FG, Wiese K, Popov AV (eds) Sensory systems and communication in arthropods. advances in life sciences, Verlag, Berlin, pp 217–228
Josephson RK, Young D (1985) A synchronous insect muscle with an operating frequency greater than 500 Hz. J Exp Biol 118:185–208

Kennedy HJ, Crawford AC, Fettiplace R (2005) Force generation by mammalian hair bundles supports a role in cochlear amplification. Nature 433:880–883
Kernan MJ (2007) Mechanotransduction and auditory transduction in *Drosophila*. Pflügers Arch 454:703–720
Kühne R (1982) Neurophysiology of the vibration sense in locusts and bushcrickets: the responses of ventral-cord neurons. J Insect Physiol 28(7):615–623
Kühne R, Silver S, Lewis B (1984) Processing of vibratory and acoustic signals by ventral cord neurones in the cricket *Gryllus campestris*. J Insect Physiol 30(7):575–585
Lane DH (1995) The recognition concept of species applied in an analysis of putative hybridization in New Zealand cicadas of the genus *Kikihia* (*Insecta*: Homoptera: *Tibicinidae*). In: Lambert DM, Spencer HG (eds) Speciation and the recognition concept: theory and application. John Hopkins University Press, Baltimore, pp 367–421
Löhe G, Kleindienst H-U (1994) The role of the medial septum in the acoustic trachea of the cricket *Gryllus bimaculatus*. II. Influence on directionality of the auditory system. J Comp Physiol A 174:601–606
Marshall DC, Cooley JR (2000) Reproductive character displacement and speciation in periodical cicadas, with description of a new species, 13-year *Magicicada neotredecim*. Evolution 54(4):1313–1325
Marshall DC, Slon K, Cooley JR, Hill KBR, Simon C (2008) Steady plio-pleistocene diversification and a 2-million-year sympatry threshold in a New Zealand cicada radiation. Mol Phylog Evol 48:1054–1066
Mason AC (1991) Hearing in a primitive ensiferan: the auditory system of *Cyphoderris monstrosa* (Orthoptera: *Haglidae*). J Comp Physio A 168:351–363
Michel K (1975) Das tympanalorgan von Cicada *orni L.* (*Cicadina*, Homoptera). Eine Licht- und Elektronenmikroskopische Untersuchung. Zoomorphologie 82:63–78
Michelsen A (1985) Time resolution in auditory systems. Springer, New York
Michelsen A (1998) Biophysics of sound localization in insects. In: Hoy RR, Popper AN, Fay RR (eds) Comparative hearing: insects. Springer, New York, pp 18–62
Michelsen A, Larsen ON (2008) Pressure difference receiving ears. Bioinsp Biomim 3(011001):18
Michelsen A, Löhe G (1995) Tuned directionality in cricket ears. Nature 375:639
Moore TE, Sawyer RT (1966) The mechanism of cicada timbal action (Insecta: Homoptera: *Cicadidae*). Am Zool 5:509
Morse PM, Ingard KU (1987) Theoretical acoustics. Princeton University Press, New Jersey 949p
Moulds MS (1990) Australian Cicadas. New South Wales University Press, New South Wales
Moulds MS (2005) Song Analyses of Cicadas of the genera aleeta moulds and tryella moulds (Hemiptera: *Cicadidae*). Proc Linn Soc New South Wales 126:133–142
Münch D (1999) Frequenz- und zeitverarbeitung durch thorakale auditorische interneurone bei zikaden (tettigetta josei). Diploma dissertation, Humboldt-University, Berlin
Myers JG (1928) The morphology of the *Cicadidae*. Proc Zool Soc Lond 25:365–472
Popov AV (1975) The structure of the tymbals and the characteristics of the sound signals in singing cicadas (Homoptera, *Cicadidae*) in the southers regions of the USSR. Entomol Obozr 54:258–291
Popov AV (1981) Sound production and hearing in the cicada *Cicadetta* sinuatipennis Osh. (Homoptera: *Cicadidae*). J Comp Physiol 142:271–280
Popov AV (1990) Co-evolution of sound-production and hearing in insects. In: Gribakin FG, Wiese K, Popov AV (eds) Sensory systems and communication in arthropods, advances in life sciences. Birkhäuser, Berlin, pp 301–304
Popov AV, Aronov IB, Sergeeva MV (1985) Calling songs and hearing in cicadas from Soviet Central Ásia. Zh Evol Biokh Fiziol 21:288–298
Pringle JWS (1954) A physiological analysis of cicada song. J Exp Biol 31:525–560
Prinz P, Ronacher B (2002) Temporal modulation transfer functions in auditory receptor fibres of the *locust* (*Locusta migratoria L.*). J Comp Physiol A 188:577–587

Richards DG, Wiley RH (1980) Reverberations and amplitude fluctuations in the propagation of sound in a forest: implications for animal communication. Am Nat 115(3):381–399
Robertson RM, Pearson KG (1982) A preparation for the intracellular analysis of neuronal activity during flight in the locust. J Comp Physiol 146:311–320
Robertson RM, Pearson KG (1984) Interneuronal organization in the flight system of the locust. J Insect Physiol 30(1):95–101
Ronacher B, Stumpner A (1988) Filtering of behaviourally relevant temporal parameters of a grasshopper's song by an auditory interneuron. J Comp Physiol A 163:517–523
Simmons PJ (1977) Neuronal generation of singing in a cicada. Nature 270:243–245
Simmons PJ, Young D (1978) The tymbal mechanism and song patterns of the bladder cicada *Cystosoma saundersii*. J Exp Biol 76:27–45
Simões PC, Quartau JA (2006) Selective responsiveness in males of Cicada *orni* to conspecific and allospecific calling songs (Hemiptera: *Cicadidae*). Entomol Gener 29(1):47–60
Stokes DR, Josephson RK (2004) Power and control muscles of cicada song: structural and contractile heterogeneity. J Comp Physiol A 190:279–290
Stölting H, Moore TE, Lakes-Harlan R (2002) Substrate vibrations during acoustic signaling in the cicada *Okanagana rimosa*. J Insect Science 2(2):1–7
Sueur J, Aubin T (2002) Acoustic communication in the palaearctic red cicada, *Tibicina* haematodes: chorus organization, calling-song structure, and signal recognition. Can J Zool 80:126–136
Sueur J, Aubin T (2003) Specificity of cicada calling songs in the genus *Tibicina* (Hemiptera: *Cicadidae*). Syst Entomol 28:481–492
Sueur J, Puissant S, Simões PC, Seabra S, Boulard M, Quartau JA (2004) Cicadas from Portugal: revised list of species with eco-ethological data (Hemiptera: *Cicadidae*). Insect Syst Evol 35:177–187
Sueur J, Windmill JFC, Robert D (2006) Tuning the drum: the mechanical basis for frequency discrimination in a mediterranean cicada. J Exp Biol 209:4115–4128
Sueur J, Vanderpool D, Simon C, Ouvrard D, Bourgoin T (2007) Molecular phylogeny of the genus *Tibicina* (Hemiptera: *Cicadicae*): rapid radiation and acoustic behaviour. Biol J Linn Soc 91:611–626
Sueur J, Windmill JFC, Robert D (2008) Sexual dimorphism in auditory mechanics: tympanal vibrations of *Cicada orni*. J Exp Biol 211:2379–2387
Sueur J, Windmill JFC, Robert D (2010) Sound emission and reception tuning in three cicada species sharing the same habitat. J Acoust Soc Am 127(3):1681–1688
Tougaard J (1996) Energy detection and temporal integration in the noctuid A1 auditory receptor. J Comp Physiol A 178:669–677
Tougaard J (1998) Detection of short pure-tone stimuli in the noctuid ear: what are temporal integration and integration time all about? J Comp Physiol A 183:563–572
Trilar T, Gogala M, Popa V (2006) Contribution to the knowledge of the singing cicadas (Auchenorrhyncha: *Cicadoidea*) of Romania. Acta Entomol Slovenica 14(2):175–182
Vogel R (1923) Ube rein tympanales sinnesorgan, das mutmassliche hororgan der singzikaden. Z Anat Entwickl-Gesh 67:190–231
von Helversen O (1979) Angeborenes erkennen akustischer schlüsselreize. Verh Dtsch Zool Ges 1979:42–59
Weber T, Moore TE, Huber F, Klein U (1988) Sound production in periodical cicadas (Homoptera: *Cicadidae*: *Magicicada septendecim*, *M. cassini*). In: Proceedings 6th auchen meeting, Turin, 7–11 Sept 1987, pp 329–336
Wohlers D, Bacon J (1980) Sexual dimorphism of motoneurons: timbal muscle innervation in male periodical cicadas and homologous structures in females. Cell Tissue Res 209:371–382
Wohlers D, Williams JDL, Huber F, Moore TE (1979) Central projections of fibers in the auditory and tensor nerves of cicadas (Homoptera: *Cicadidae*). Cell Tissue Res 203:35–51
Young D (1972a) Neuromuscular Mechanism of Sound Production in Australian Cicadas. J Comp Physiol 79:343–362

Young D (1972b) Analysis of songs of some Australian cicadas (Homoptera: *Cicadidae*). J Aust Ent Soc 11:237–243
Young D (1990) Do cicadas radiate sound through their ear-drums? J Exp Biol 151:41–56
Young D, Bennet-Clark HC (1995) The role of the tymbal in cicada sound production. J Exp Biol 198:1001–1019
Young D, Hill KG (1977) Structure and function of the auditory system of the cicada, *Cystosoma saundersii*. J Comp Physiol 117:23–45
Young D, Josephson RK (1983a) Mechanisms of sound-production and muscle contraction kinetics in cicadas. J Comp Physiol 152:183–195
Young D, Josephson RK (1983b) Pure-tone songs in cicadas with special reference to the genus *Magicicada*. J Comp Physiol 152:197–207

Chapter 8
Towards an Understanding of the Neural Basis of Acoustic Communication in Crickets

Berthold Hedwig

Abstract Their conspicuous acoustic communication behaviour makes crickets excellent model systems to study the neural mechanisms underlying signal generation and auditory pattern recognition. Male singing is driven by a central pattern generator (CPG) housed in the metathoracic and anterior abdominal ganglia with rhythmically active opener and closer interneurons that can reset the chirp rhythm. Command neurons descending from the brain control the singing behaviour. Female phonotaxis is tuned towards the species-specific pattern of the male calling song and auditory orientation behaviour demonstrates a parallel organisation of pattern recognition and highly accurate steering. First order auditory processing occurs in the thorax and pattern recognition in the brain. Local auditory brain neurons are tuned to the structure of the calling song, based on fast integration of inhibitory and excitatory synaptic activity. How pattern recognition is linked to the generation of auditory steering commands still remains an open question.

8.1 Introduction

More than 2,000 species of crickets use species-specific acoustic signals for intraspecific communication (Alexander 1962). Today, their conspicuous behaviour has been the focus of ethological, biophysical and neurophysiological research for 100 years (Regen 1913) and has led to major scientific contributions to the comprehensive characterisation of an insect auditory behaviour and its neural pathway (Huber and Thorson 1985), the neural basis of bat avoidance (Nolen and Hoy 1984), neural mechanisms of selective attention (Pollack 1988; Sobel and Tank 1994) the cellular mechanisms of a corollary discharge mechanism

B. Hedwig (✉)
Department of Zoology, University of Cambridge, Cambridge CB2 3EJ, UK
e-mail: bh202@cam.ac.uk

B. Hedwig (ed.), *Insect Hearing and Acoustic Communication*,
Animal Signals and Communication 1, DOI: 10.1007/978-3-642-40462-7_8,

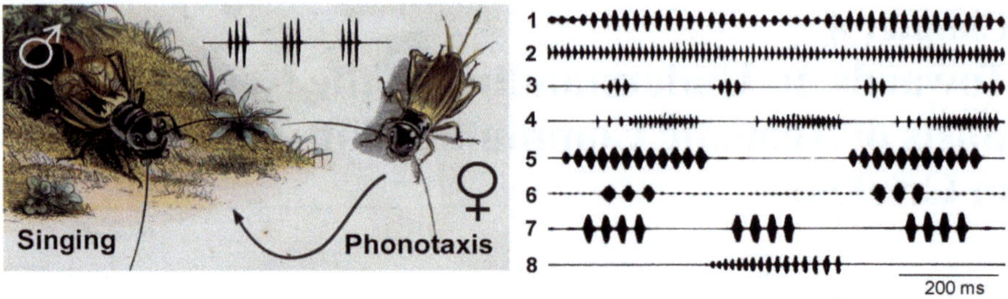

Fig. 8.1 **a** Male and female cricket acoustic communication behaviour; while the male sings its calling song in front of a burrow the female approaches by orienting towards the song. Modified from Roesel von Rosenhof (1749) species probably *G. campestris.* **b** Variety of calling song patterns of sympatric cricket species in Azerbaijan. (1) *Pteronemobius heydeni* Fish., (2) *Tartarogryllus bucharicus* B.-B.; (3) *Tartarogryllus tartarus obscurior* Uv. (4) *Tartarogryllus burdigalensis* Latr. (5) *Oecanthus pellucens* Scop. (6) *Turanogryllus lateralis* Fieb. (7) *Gryllus bimaculatus* DeGeer (8) *Modicogryllus pallipalpis* Farb. Modified from (Popov et al. 1974; Fig. 8.1) with permission of the Nordrhein-Westfälische Akademie der Wissenschaften und der Künste

(Poulet and Hedwig 2002, 2006), biophysical and neural mechanisms of directional hearing (Michelsen et al. 1994; Michelsen and Löhe 1995; Schöneich and Hedwig 2010) frequency processing in the afferent and central pathway (Nocke 1972; Pollack and Faulkes 1998) and recently to implementing these findings on auditory orientation in bio-inspired robots (Reeve and Webb 2003). This paper will review aspects of these advances related to the neural mechanisms underlying male singing and female phonotactic behaviour.

8.2 Male Singing Behaviour

Male crickets produce a calling, rivalry or courtship song by rhythmically rubbing a plectrum at the edge of the left front wing against a file at the underside of the right wing. The underlying opening and closing wing movements are powered by muscles driven by motoneurons in the meso- and also the prothoracic ganglion (Kutsch 1969). Only closing movements generate a sound pulse setting structures on the wings in resonant coherent oscillations (Montealegre-Z et al. 2011). A salient feature of cricket auditory behaviour is the broad variety of species-specific song patterns, encompassing pulsing, chirping and trilling calling songs (Popov et al. 1974; Fig. 8.1). In some species song structure is the only feature to discriminate between identical morphs and points towards evolutionary changes in the neural mechanisms controlling singing as decisive steps in speciation (Otte 1992).

8.2.1 Neuropharmacology and Fictive Singing Pattern

Focal electrical brain stimulation in *Gryllus campestris* and *Gryllus bimaculatus* demonstrated the importance of the brain for singing behaviour and especially that the anterior protocerebrum controls the generation of calling song (Kutsch and Huber 1989). Mircoinjection of neuroactive substances in this brain area allowed systematic testing of neuroactive substances that release singing behaviour (Otto 1978; Wenzel and Hedwig 1999). In intact tethered males with only the head capsule opened, injections of cholinergic agonists (0.5 nL, 10^{-3} mol/l^{-1}) were effective to elicit singing (Fig. 8.2a). Activation of ligand coupled nicotinic ACh-recepetors with acetylcholine or nicotine had fast effects and elicited bouts of singing after a latency of 11.5 s. Injection of muscarine, which activates a second messenger cascade via muscarinic ACh receptors, led to a gradual build-up of the behaviour over 60 s with singing sequences lasting for several minutes. Subsequent injection of GABA stopped ongoing singing activity. The experiments point towards a cholinergic pathway driving the release of singing. This pathway can effectively be activated by injection of Eserine, an acetylcholine-esterase inhibitor, causing a gradual build-up of ACh in the tissue. After injection of Eserine, singing can take several minutes to start, but then may continue for several hours. For intracellular studies of the cricket CNS pharmacological brain

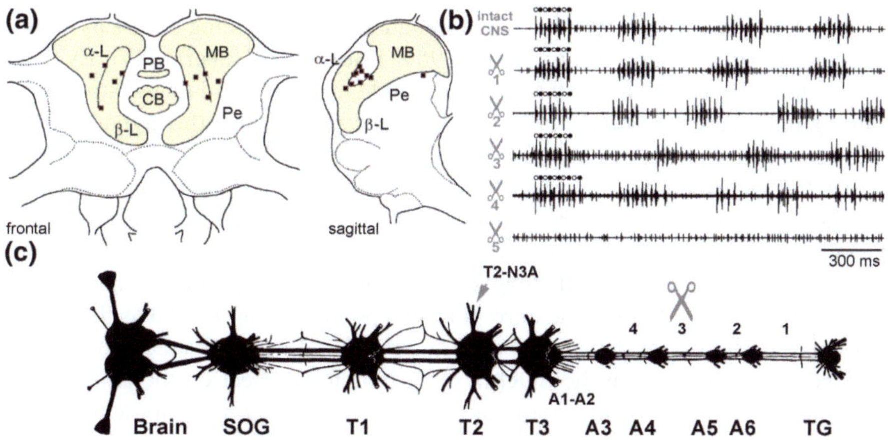

Fig. 8.2 **a** Brain of *G. bimaculatus*; micro-injection sites (■) eliciting singing behaviour are located between the α-lobe and the pedunculus. **b**, **c** Fictive singing motor pattern and cricket CNS. During successive sections of abdominal connectives singing always stopped when the T3–A3 connectives were severed. Motor pattern recorded from wing nerve T2-N3A, opener and closer motoneuron activity indicated by (○) and (●). MB mushroom body, α-L alpha lobe, β-L beta lobe, Pe pedunculus, CB central body complex, PB protocerebral bridge. **a** From Wenzel and Hedwig (1999), **b** from Schöneich and Hedwig (2011), **c** modified from (Huber 1955; Fig. 1) with permission of John Wiley and Sons

stimulation with Eserine is an efficient way to release singing activity, even after all thoracic sensory and motor nerves are cut a fictive singing motor pattern is generated (Poulet and Hedwig 2006).

Fictive singing motor activity can be recorded as spike activity of wing-opener and closer motoneurons in the mesothoracic nerve 3A (Fig. 8.2b) and closely resembles the timing of the normal pattern in *G. bimaculatus* (Schöneich and Hedwig 2011, 2012). As in normal singing the mean chirp rate ranges from 2 to 3 Hz and the mean pulse rate is about 24 Hz. The chirp rate decreases when the pulse number per chirp increases from 3 to 5. In contrast, the chirp interval always remains stable between 210 and 260 ms. Over a chirp the mean opener–closer interval is very constant and between 21 and 24 ms for different males, whereas the closer–opener interval may increase from about 19–25 ms causing a gradual increase in the pulse-period. The close similarity between fictive and normal singing motor pattern indicates that sensory feedback is not required for generating the singing pattern; it might however, be used during sound production for fine adjustments of the front wing movements (Kutsch and Huber 1989).

8.2.2 Locating the Singing CPG

As the singing wing movements are mainly controlled by motoneurons in the mesothoracic ganglion T2, it was assumed that the singing CPG is also located in the thoracic ganglia with the brain and the reproductive organs being necessary to trigger calling and courtship song (Huber 1955; Kutsch and Huber 1989). Evidence based on micro-lesions in the CNS (Hennig and Otto 1996) and on the structure of singing interneurons (Hennig 1990) gradually pointed against this hypothesis and towards a contribution of abdominal ganglia to singing pattern generation. When connectives between the abdominal ganglion A3 and the terminal ganglion TG are cut during fictive singing there is little effect on singing activity. However, singing always immediately stops when connectives between T3 and A3 are sectioned (Fig. 8.2b, c; Schöneich and Hedwig 2011). Thus, a crucial part of the singing network must be housed in A3 and current results point towards T3–A3 as the ganglia housing the network of the singing pattern generator. Beyond this descending ventilatory interneurons from the SOG have an inhibitory effect on the chirp pattern (Otto and Hennig 1993) and coordinate ventilatory and singing motor activity.

8.2.3 Unravelling the CPG Network for Singing

Microelectrode recordings within the singing network revealed two crucial opener interneurons (Schöneich and Hedwig 2012), which are activated just before the opener motoneurons. The metathoracic-descending-opener interneuron has dendrites running at the midline posteriorly towards the abdominal neuromeres A1 and A2 which are fused with T3 (Fig. 8.3a). Its axon projects into the abdominal nerve cord giving off characteristic anterior and posterior axonal collaterals in ganglia A3–A6. There is only one of these interneurons at each side of the CNS. Activity in this interneuron is characterised by typical rhythmic membrane potential oscillations with consecutive de- and hyperpolarisation of the dendritic membrane potential in the pulse pattern and burst of spikes generated during the depolarisation phase (Fig. 8.3b). Manipulating spike activity of this interneuron by intracellular current injection requires unusual high currents of 10–20 nA, but if successful, elicits additional sequences of open-closer activity during fictive singing that reset the chirp rhythm.

An ascending opener interneuron is located in A3 where its dendrites closely match the axonal arborisation pattern of the metathoracic-opener interneuron (Fig. 8.3c). The axon of the abdominal-opener interneurons runs up to T1 with axon collaterals projecting towards the midline of the thoracic ganglia. There is a left–right pair of abdominal-opener interneurons, which may be electrically coupled by gap-junctions as dye-coupling is observed in neurobiotin staining. Also, the abdominal-opener interneuron generates rhythmic membrane potential

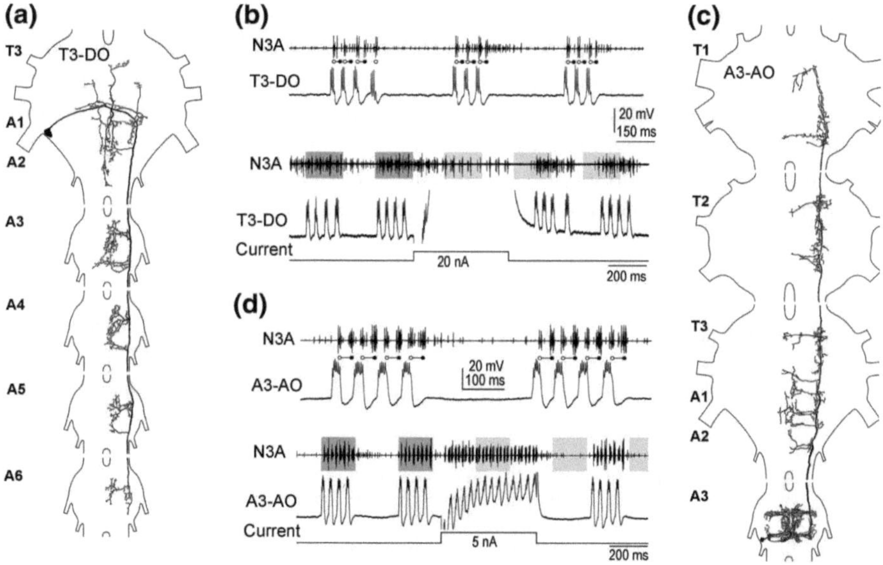

Fig. 8.3 **a** Structure of the metathoracic opener interneuron T3–DO descending from T3 towards A6. **b** A burst of spikes in the interneuron precedes the opener phase of singing activity and inhibition precedes the closer activity. Depolarisation of the interneuron elicits a bout of opener–closer activity and resets the chirp pattern. **c** Structure of the abdominal-opener interneuron A3–AO ascending from A3 towards T1. **d** Activity of the interneuron is coupled to the opener activity; a depolarising current pulse elicits a sequence of opener–closer activity and resets the chirp pattern. Expected timing of undisturbed chirp pattern indicated by light grey shades; modified from Schöneich and Hedwig (2012)

oscillations in phase with the opener motor activity; bursts of spikes are generated during each depolarisation, which are terminated by an inhibition (Fig. 8.3d). Injection of depolarising current pulses reliably elicits complete opener–closer cycles for the duration of current injection and reliably resets the chirp pattern, as the chirp interval between the last syllable of the elicited chirp and the first syllable of the subsequently generated chirp remains constant. Injection of current pulses within a chirp has no effect. Extended current injection even drives the CPG to produce long-lasting chirps, controlled by the duration of the current pulse. Thus, activity of the abdominal-opener interneuron not only controls the chirp rhythm but also chirp duration. Activity in the abdominal-opener interneuron starts 10 ms before the opener activity in the wing nerve and 3 ms before the metathoracic-opener interneuron, and so far appears to be the first interneuron activated in the syllable cycle.

Closer interneurons have been recorded in the anterior abdominal neuromers A1–A3 (Schöneich and Hedwig 2012) but have not yet been anatomically identified (Fig. 8.4a). They spike in phase with the wing-closer motoneurons and are inhibited during the wing-opening phase. Furthermore, over the course of a singing episode they receive a gradual inhibition, shifting their membrane potential below the resting potential. Depolarising current pulses also reset the chirp pattern. Some closer interneurons receive a barrage of IPSPs before the start of any singing sequence, which are coupled to subsequent depolarisation. Also, hyperpolarising current injection in closer interneurons is reliably followed by subsequent depolarisation, which increases with the amplitude of hyperpolarisation. Such responses of the membrane potential point towards a post-inhibitory rebound mechanism in the closer interneurons, which may be crucial for singing motor pattern generation.

A simple tentative circuit for the singing CPG can be based on reciprocally inhibitory coupled opener and closer interneurons and a post-inhibitory rebound

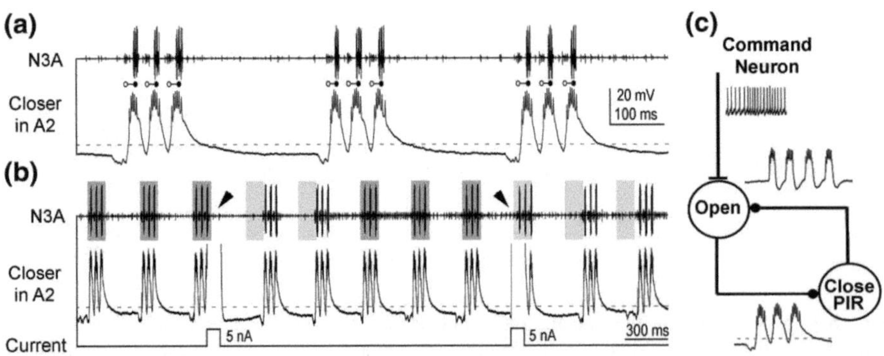

Fig. 8.4 **a** An unidentified closer interneuron receives inhibition in the opener phase and is depolarised in the closer phase. **b** Depolarising current pulses reset the chirp pattern, but do not elicit additional singing motor activity. **c** A speculative model of the singing CPG network based on mutual inhibition, post-inhibitory rebound and excitatory drive from command neurons may be sufficient to generate the pulse pattern; **a, b** from Schöneich and Hedwig (2012)

mechanism, which are sufficient to generate rhythmic neuronal activity (Fig. 8.4c; Bentley 1969; Perkel and Mulloney 1974). Tonic activity of descending command neurons (see below) may drive the opener interneurons, which generate a burst of spike activity and inhibit the closer interneurons. Due to post-inhibitory rebound the closer interneurons depolarise, generating a burst of spikes and inhibiting the opener neurons during the closer cycle. The excitation from the descending command neurons starts the opener activity again and the chirp pattern continues until some intrinsic properties of the opener interneurons reduce their activity and terminate the chirp. Singing stops, when the command activity comes to an end. As the singing circuit requires no sensory feedback, just changing the intrinsic properties of the CPG neurons or the strength of synaptic coupling could be an effective way to generate species-specific singing pattern.

As the layout of the auditory pathway is highly conserved in different species of crickets and bush-crickets (Stumpner and von Helversen 2001) we may assume a similarly conservative consistency at the pattern generation side. Thus the CPG interneurons identified in *G. bimaculatus* may provide the blueprint for the organisation of the singing motor network and comparative studies may now demonstrate how singing circuits are adapted for species-specific signalling.

8.2.4 Brain Neurons Controlling Calling Song

Extracellular electrical stimulation of the neck connectives (Otto 1971) or of small fibre bundles in the cervical connectives is sufficient to elicit singing (Bentley 1977), and indicates the existence of descending singing command neurons (Kupfermann and Weiss 1978). Probing the crucial regions of the anterior protocerebrum with intracellular recordings demonstrated a calling song command neuron at each side of the brain (Fig. 8.5a, Hedwig 2000). The primary neurite of this neuron runs towards the dorsally located cell body and dendrites arborise in the frontal medial area of the protocerebrum. Dendrites are densely packed within a delimited neuropil area between the pedunculus and the α-lobe of the mushroom body. The axon of the neuron descends in the contralateral connective towards the thoracic ganglia; its arborisation pattern in the ventral nerve cord however is not known.

This neuron fulfils the criteria set up for the characterisation of command neurons (Kupfermann and Weiss 1978). When its spike activity is driven to 100–120 AP/s by intracellular current injection into its dendrites, crickets will lift their wings and start singing (Fig. 8.5b). Singing is maintained while the tonic activity of the neuron is kept at about 120 AP/s. With the end of current injection and a decrease in the command spike activity singing ceases within a few seconds. Singing activity generally is facilitated with repetitive sequences of current injection, to the extent that even short 1 s current pulses can trigger ongoing singing, while the command neuron activity then is as low as 30–35 AP/s. Stimulation of the command neurons can alter the overall chirp rate but it has no effect on the syllable repetition rate (Fig. 8.5c; Otto 1971; Bentley 1977, Hedwig 2000). Syllable

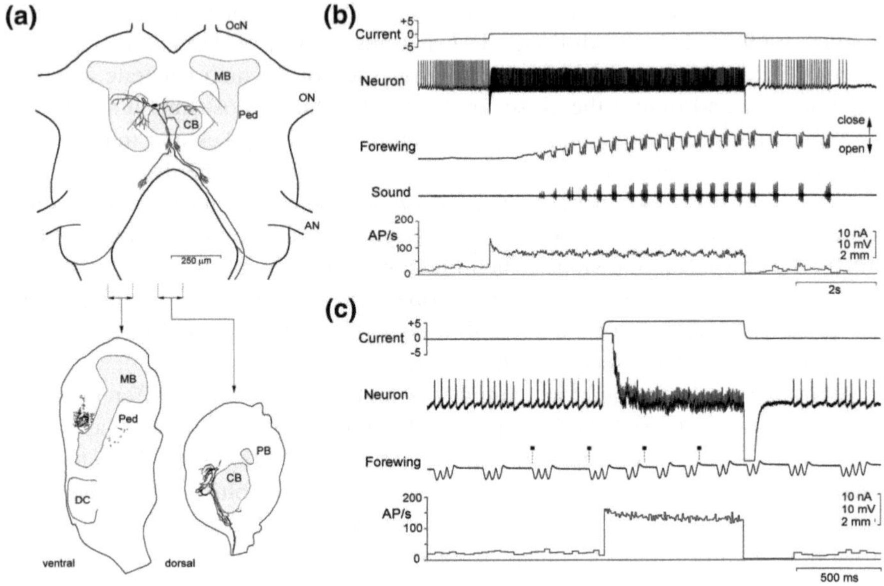

Fig. 8.5 **a** Structure of the descending calling song command neuron in *G. bimaculatus*, showing dense dendritic arborisations occurring between the α-lobe and the pedunculus. ON optic nerve, OcN ocellar nerve, Dc Deutocerebrum, see Fig. 8.2 for other abbreviations. **b** Current injection in the interneuron in an intact tethered male elicits a spike rate of about 120 AP/s and singing, that gradually wanes after current injection. **c** Current injection during singing accelerates the chirp pattern and causes a reduction in the number of pulses per chirp. Singing activity monitored by wing movements and sound pattern; modified from Hedwig (2000)

generation therefore appears to be largely independent of descending control. In intracellular stimulation experiments the command neuron always only elicited calling song, whereas in extracellular brain and connective stimulation experiments transitions between different songs occurred. Therefore, clear-cut conclusions may not yet be drawn on the control of different song patterns in crickets, however, at least in grasshoppers parallel descending pathways are required to control the full spectrum of male singing activity (Hedwig and Heinrich 1997).

8.3 Female Phonotaxis and Peripheral Auditory Pathway

Female crickets walk or fly towards singing conspecific males using acoustic cues for orientation (Regen 1913; Ulagaraj and Walker 1973). Their ears provide an excellent opportunity to study how insect hearing organs operate within the limitations set by the laws of physics (Michelsen 1992) and how the neural auditory pathway processes directional information and species-specific song patterns as the basis of phonotactic behaviour (Popov et al. 1974; Hoy 1978; Huber and Thorson 1985; Pollack 2000). A hearing organ is located in each front tibia comprising an array of 40–60 auditory afferents linearly arranged in the *crista acoustica*. Sound enters the auditory system via two tympanic membranes and an auditory trachea, with openings in the first thoracic segment that connects both ears. The structure (Ball et al. 1989), biophysics (Larsen et al. 1989) and afferent activity (Esch et al. 1980; Imaizumi and Pollack 2001) have been analysed in detail and appear as specialisations for directional hearing and for low and high frequency discrimination. It is still unknown which forces and structures trigger the mechano-electrical transduction process in afferent neurons and how their frequency tuning is established.

8.3.1 Temporal Tuning of Phonotaxis

Behavioural studies on phonotaxis have focussed on auditory pattern recognition and on the mechanisms of auditory orientation using arena experiments (Murphey and Zaretsky 1972; Tschuch 1977; Rheinlaender and Blätgen 1982; Stout et al. 1983), flight paradigms (Pollack and Hoy 1979) or treadmill systems (Wendler et al. 1980; Thorson et al. 1982; Doherty and Pires 1987; Hedwig and Poulet 2004) the latter of which monitor the animals' walking speed and direction and indicate its phonotactic response (Fig. 8.6a). When exposed to acoustic patterns with systematic variations of the pulse duration and pulse interval, in all test paradigms female crickets prefer temporal patterns that match the species-specific male calling song (Fig. 8.6a). Although the pulse period may be a crucial component for phonotactic orientation in *G. campestris and G. bimaculatus* (Thorson et al. 1982), other parameters like pulse number and chirp rate contribute to the phonotactic response as well (Stout et al. 1983). The response to auditory patterns is not fixed, but transiently modulated by phonotaxis itself. If non-attractive chirp patterns are inserted into a sequence of normal calling song, the animals will steer towards these non-attractive chirps as well, which do not elicit phonotaxis when presented on their own (Fig. 8.6b; Poulet and Hedwig 2005). In this way phonotaxis broadens the acceptance of auditory patterns and the animals transiently tolerate distorted calling song patterns as they occur in natural environments due to sound diffraction and noise (Kostarakos and Römer 2010). The system may operate to maximise phonotactic targeting success while pursuing conspecific auditory patterns.

8.3.2 Auditory Steering and Pattern Recognition

How do pattern recognition and directional steering interact during auditory orientation? Are they organised in serial or parallel pathways (Helversen and Helversen 1995)? Early trackball recordings of female orientation apparently showed that steering responses only occur after processing of complete chirps (Schildberger

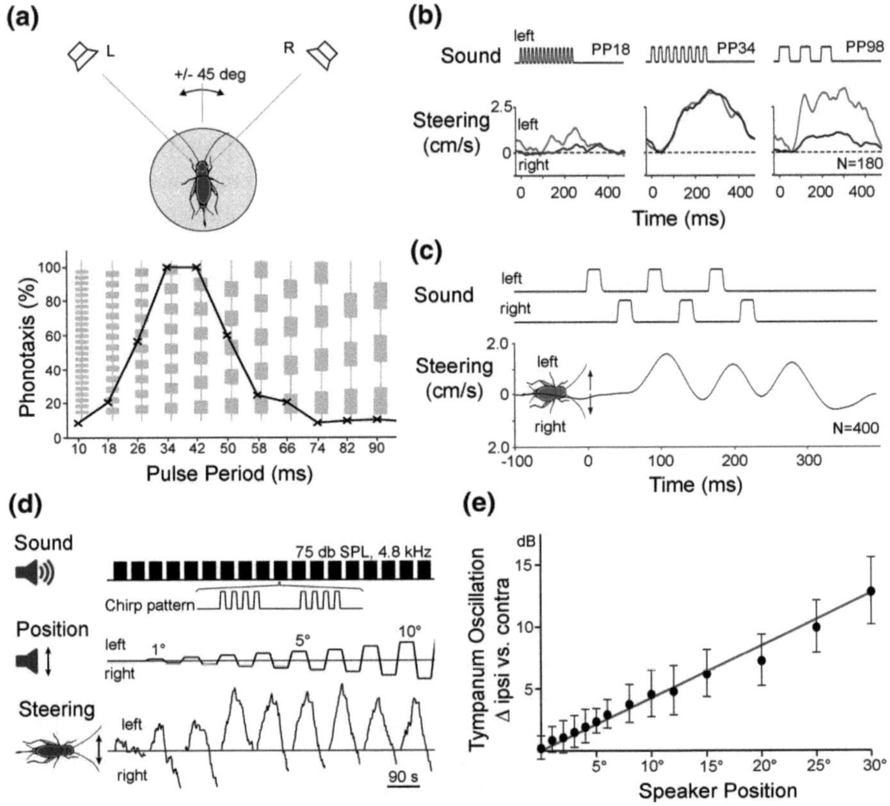

Fig. 8.6 **a** Open-loop track ball system for analysing cricket phonotactic behaviour. While the tethered female walks on the trackball different sound patterns are presented from left or right and the movement of the trackball is monitored. Female phonotaxis of *G. bimaculatus* is tuned to the temporal pattern of the species-specific calling song. **b** Modulation of phonotactic tuning. Steering responses to different chirps during phonotaxis (*black lines*). Responses to pulse periods PP18 and PP98 (*grey line*) are stronger when these are inserted into a sequence of normal calling song. **c** Rapid steering responses to individual sound pulses occur during presentation of a split calling song pattern. **d** Accuracy of directional steering allows crickets to reliably steer to speaker positions 1–2° off their length axis. **e** The bilateral difference in tympanic membrane vibrations corresponds to 0.4 dB/° **a** from Hedwig (2006), **b**, **c** from Poulet and Hedwig (2005), **d**, **e** from Schöneich and Hedwig (2010)

and Hörner 1988; Schmitz et al. 1982) and indicated a serial organisation of pattern recognition and steering. Auditory steering responses are, however, surprisingly fast when monitored with a low inertial track ball system (Fig. 8.6c; Hedwig and Poulet 2004, 2005). Females exposed to split-song paradigms of the calling song, with subsequent sound pulses presented from opposite sides, steer towards individual pulses with a latency of just 55–60 ms. As the animals do not evaluate a complete chirp to guide their orientation, auditory pattern recognition and steering have to be organised in parallel pathways; pattern recognition however is first required for fast steering responses to occur. Thus pattern recognition and steering are transiently processed in a serial manner and once recognition is established they then occur in parallel. Overall female phonotaxis is characterised by three properties: it is tuned towards the species-specific sound pattern, steering is fast and precise and steering responses are transiently modulated.

8.3.3 Directional Hearing and Orientation

Phonotactic orientation requires directional auditory sensitivity and processing. The primary auditory afferents project into the prothoracic ganglion which is the first stage of auditory processing (Esch et al. 1980). The directionality of the system appears to be tuned to the carrier frequency of the calling song (Michelsen and Löhe 1995). Due to phase shifts in the auditory trachea the cricket's hearing system generates bilateral response differences up to 20–30 dB. Directional cues are, however, small in the frontal range of the animal as the overall gradient for directional discrimination may only be 10–12 dB over 25° and appear to be even less for small angular deviation. This led to the assumption that crickets face a frontal area of uncertainty where they cannot consistently turn towards a sound (Rheinlaender and Blätgen 1982; Larsen et al. 1989), in line with reports that the animals use a meandering walk when approaching a singing male (Schmitz et al. 1982). When phonotactically walking females were tested on an open-loop trackball with with a constant alignment of their body relative to the sound signal, they showed a remarkably precise directional orientation and reliably steered towards calling songs, which were just 1° off their length axis (Fig. 8.6d, e; Schöneich and Hedwig 2010). Relating behavioural directional discrimination to bilateral response differences of tympanic membrane oscillations and auditory afferents revealed that the animals can use a slope of 0.4 dB/° interaural signal difference in the frontal ±30° for their directed phonotactic approach. In summed recordings of the afferent response this gradient is reflected in changes of response latency of 42 μs/°. Interaural amplitude and latency differences of the afferent activity are forwarded to the thoracic auditory neuropil where reciprocal inhibition by the local omega interneurons (ON1) allows for further bilateral contrast enhancement leading to highly directional responses of the ON1 neurons even at small angles close to frontal stimulation (Wiese and Eilts-Grimm 1985; Wohlers and Huber 1982).

On each side of the CNS only two auditory interneurons forward auditory activity towards the brain: interneuron AN1 is tuned in the range of 4–5 kHz the carrier frequency of the calling song and AN2 responds best to high-frequency sounds in the range of 10–15 kHz. The activity of AN1 is crucial for phonotactic steering (Schildberger and Hörner 1988), whereas AN2 triggers bat-avoidance behaviour (Nolen and Hoy 1984). When females steer towards a sound pattern presented from the left and the left AN1 (i.e., axon in the left connectives) is hyperpolarised to reduce its spike activity, then these animals alter their direction and walk towards the right (Schildberger and Hörner 1988). The conclusion from these experiments was that the animals steer towards the side where the calling song activates AN1 more strongly, by comparing bilateral AN1 activity levels in the brain. To meet the requirements of reflex-like fast steering, descending motor commands could be generated in the brain. However, as pattern recognition modulates the steering process, descending interneurons might just alter the response properties of the thoracic sensory-motor pathway involved in the control of walking (Poulet and Hedwig 2005). Overall in terms of directional hearing, crickets together with the fly *Ormia ochracea* (Mason et al. 2001), are among the most precise invertebrates and can easily compete with vertebrate achievements (Heffner et al. 2007).

8.3.4 Auditory Brain Neurons and Pattern Recognition

Crickets have been established as a model system for the processing and recognition of acoustic pulse patterns. No temporal filtering occurs at the level of the thoracic auditory processing, AN1 rather copies any auditory patterns presented at the species-specific carrier frequency (Schildberger et al. 1989). How is phonotaxis controlled by the brain? Different concepts on neural mechanisms underlying pattern recognition have been put forward such as feature detection via cross-correlation, autocorrelation via a delay line and a combination of high- and low-pass filters (see Weber and Thorson 1989).

What is the evidence from brain neurons? Based on the response properties of local auditory brain neurons Schildberger (1984) concluded that *G. bimaculatus* the band-pass like tuning of female phonotaxis is due to specific low-pass and high-pass filter neurons, which are finally combined to result in band-pass tuned neural responses, akin to auditory filtering in the anuran brain (Rose and Capranica 1983). This concept is challenged by recent recordings from local auditory brain neurons, in which neural response towards different pulse patterns were compared with phonotactic responses to corresponding sound patterns (Kostarakos and Hedwig 2012). The brain neurons closely match the axonal arborisation pattern of AN1 in the anterior protocerebrum (Fig. 8.7a). The axonal projections of AN1 as well as the neurites of the interneurons form a ring-like arborisation pattern in the anterior ventral protocerebrum. Although the structure of the local auditory brain neurons is similar, their auditory response properties are very different and demonstrate an increasing level of tuning towards temporal patterns (Fig. 8.7b, c). When tested with different pulse periods female phonotaxis is tuned towards pulse periods of 34–42 ms. The B-LI2 neuron responds phasically with EPSPs and spikes to the individual sound pulses of all patterns. Its activity even reflects minor differences in the overall sound energy presented. Its response function, however, exhibits no tuning to the different pulse patterns and like the ascending interneuron AN1 its does not match the phonotactic behaviour. At the level of B-LI3 a more complex tuned response pattern is apparent, as the strongest excitatory response occurs to the second pulse of a chirp. Moreover, the response of B-LI3 to short and long pulse periods is about 60 %

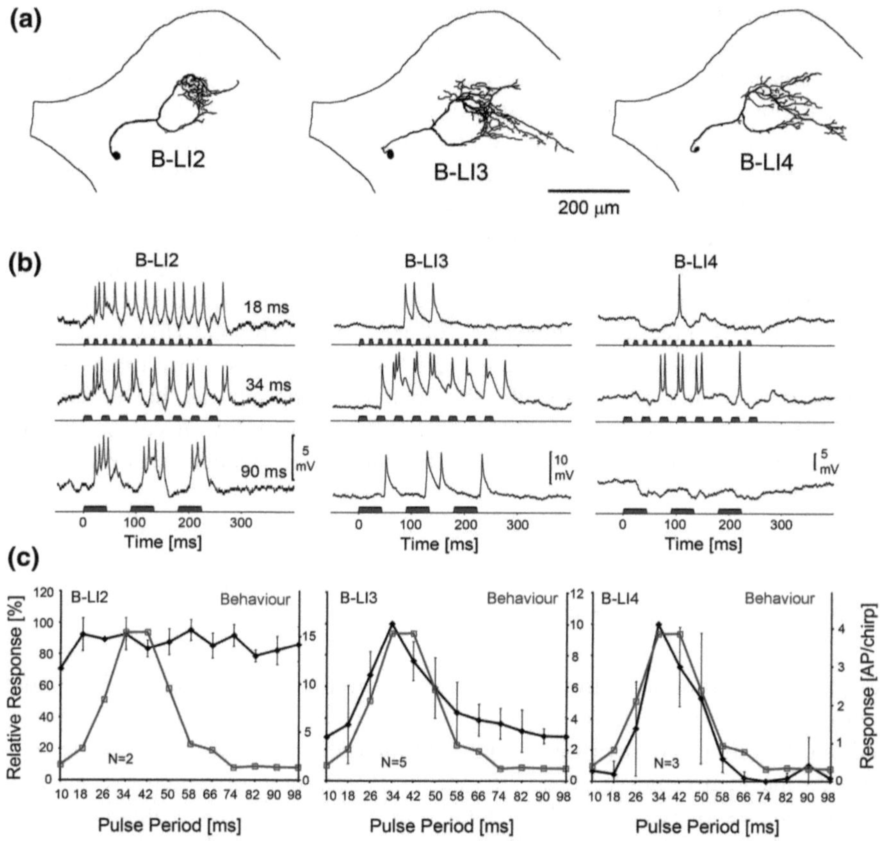

Fig. 8.7 **a** Structure of local auditory brain neurons in *G. bimaculatus* in the anterior protocerebrum, close to the axonal arborisation of the ascending interneuron AN1. **b** The neurons exhibit different degrees of temporal filtering when exposed to sound pattern with different pulse periods. Activity pattern in B-LI4 is shaped by inhibitory and excitatory synaptic inputs. **c** Whereas B-LI2 shows no sign of filtering, spike activity in B-LI3 and especially B-LI4 closely match the tuning of female phonotactic behaviour; modified from Kostarakos and Hedwig (2012)

lower than the response to chirps presented at the species-specific pulse period. Tuning to the species-specific temporal pattern is best in B-LI4 neurons, which receive a mixture of inhibitory and excitatory synaptic inputs. Inhibition dominates the neuron's response, when chirps with low or high pulse periods are presented. At the species-specific pulse period, however, the excitatory spiking response of the neuron is strongest leading to a close match between the tuning of the neural response and phonotactic behaviour.

While the temporal tuning of the local auditory brain neurons becomes sharper from B-LI2 to B-LI4, their spike response latency increases from 21 to 37 ms and their maximum response decreases from 14.3 to 3.9 AP/Chirp, respectively. This is in line with the concept of sparse coding, shifting the representation of stimulus features from a temporal activity-based code to a neuron-specific place code, which appears to be an energetically efficient way for small nervous systems to ensure a robust representation of stimulus patterns (Olshausen and Field 2004).

Thus, in the cricket brain temporal processing and tuning to the species-specific pattern seems to occur in a network of local auditory interneurons, closely matching the axonal arborisation pattern of AN1. Auditory band-pass properties encountered in other brain interneurons may just be the consequence of this processing. Like in frogs (Edwards et al. 2007; Rose et al. 2011) auditory processing appears to be based on fast pulse-by-pulse interactions of excitatory and inhibitory synaptic activity; the precise mechanisms still need to be explored. How many tuned filters for pulse patterns may crickets have? As the local auditory brain neurons are only tuned to the species-specific song pattern, it appears that they employ just one filter for pattern recognition. This is besides their ability of categorical discrimination of sounds presented with different carrier frequencies (Wyttenbach et al. 1996).

8.3.5 The Auditory-to-Motor Loop

It is not yet clear how auditory pattern recognition is linked to descending motor control for phonotaxis. Two types of motor activities are required: locomotion and auditory-induced steering. About 200 interneurons descend from the brain towards the ventral nerve cord (Staudacher 1998) some of which exhibit auditory responses in resting and especially in walking crickets (Staudacher and Schildberger 1998; Staudacher 2001; Zorovic and Hedwig 2011). However, due to response latencies and/or variability in their spike patterns the properties of these descending neurons are not sufficient to drive the fast phonotactic steering responses. As phonotactic steering occurs already to the very first pulse of a chirp with a latency of only 55–60 ms, pattern recognition cannot directly be involved in the steering process. However, pattern recognition apparently modulates auditory steering, as the animals respond to non-attractive sounds inserted in or presented immediately after a sequence of calling song (Poulet and Hedwig 2005). The underlying modulatory mechanism may be central to complete the auditory-to-motor loop. It may be elucidated by pharmacological approaches and by recording descending pre-motor brain neurons during phonotactic behaviour.

At the thoracic level the integration of phonotactic steering responses into the cricket's walking pattern has been elucidated by identifying the motoneurons (Baden and Hedwig 2008) and by high speed video recordings (Witney and Hedwig 2011; Petrou and Webb 2012). During phonotactic walking, while exposed to the calling song, females do not couple their stepping cycle to the chirp pattern. They rather integrate phonotactic steering into the ongoing walking pattern and alter the trajectories of the legs to accommodate changes in walking

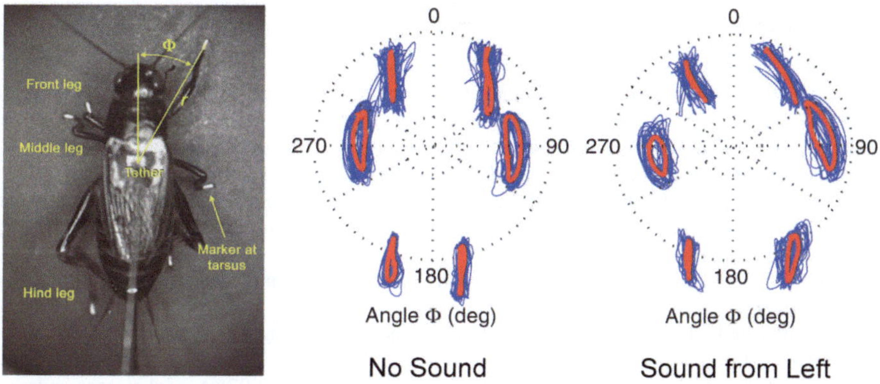

Fig. 8.8 **a** Monitoring leg movements and tarsal positions in phonotactic walking crickets with high speed video recordings. Tarsi are marked with a white spot and the angle between tarsi and the animal's length axis is measured. **b** Tarsal trajectories in a cricket walking straight ahead without acoustic stimulation. Upon orienting towards a calling song pattern from the *left* the tarsal trajectories tilt to the *left*. Radius of polar-plot 1.5 cm. Modified from Witney and Hedwig (2011)

direction. Although the trajectories of all legs are altered (Fig. 8.8), changes in the front leg movements appear to be most prominent for auditory steering and are accompanied by bending the first thoracic segment against the thoracic box, supporting steering towards the sound source.

8.3.6 Conclusion: Crickets as a Neurobiological Model System for Acoustic Communication

Exploring the cricket nervous system with intracellular recordings has allowed us to establish a close link between neural pathways, identified neurons, neural mechanisms and auditory behaviour. Future experiments may draw upon the variety of song patterns in different species and provide insight into how species-specific neural networks for auditory communication were shaped during evolution and to what degree these networks develop in a sex-specific way. However, supplementary to a neurophysiological approach cricket neurobiology may advance by genetic techniques that allow expressing genetically encoded calcium indicators in the central nervous system or subset of cell lines. As first successful steps to sequence the cricket genome are being made (Danley et al. 2007) and first transgenic crickets have been created (Nakamura et al. 2010) the future looks promising.

Acknowledgments I am grateful to the members of the lab that contributed to the presented data and fruitful discussions: Stefan Schöneich, Kostas Kostarakos, Hannah ter Hofstede, James Poulet, Tom Baden and Alice Witney and also to the BBSRC, the Isaac Newton Trust and the Royal Society for funding.

References

Alexander RD (1962) Evolutionary change in cricket acoustical communication. Evolution 16:443–467

Baden T, Hedwig B (2008) Front leg movements and tibial motoneurons underlying auditory steering in the cricket (*Gryllus bimaculatus deGeer*). J Exp Biol 211: 2123–2133

Ball E, Oldfield B, Rudolph K (1989) Auditory organ structure, development, and function. In: Huber F, Moore TE, Loher W (eds) Cricket behaviour and neurobiology. Cornell University Press, London, pp 391–422

Bentley D (1977) Control of cricket song patterns by descending interneurons. J Comp Physiol A 116:19–38

Bentley DR (1969) Intracellular activity in cricket neurons during the generation of behaviour patterns. J Insect Physiol 15:677–684

Danley P, Mullen S, Liu F, Nene V, Quackenbush J, Shaw KL (2007) A cricket gene index: a genomic resource for studying neurobiology, speciation, and molecular evolution. BMC Genomics 8:109. doi:10.1186/1471-2164-8-109

Doherty JA, Pires A (1987) A new microcomputer-based method for measuring walking phonotaxis in field crickets (Gryllidae). J Exp Biol 130:425–432

Edwards CJ, Leary CJ, Rose GJ (2007) Counting on inhibition and rate-dependent excitation in the auditory system. J Neurosci 27:13384–13392

Esch H, Huber F, Wohlers DW (1980) Primary auditory neurons in crickets: physiology and central projections. J Comp Physiol A 137:27–38

Hedwig B (2000) Control of cricket stridulation by a command neuron: efficacy depends on the behavioral state. J Neurophysiol 83:712–722

Hedwig B, Heinrich R (1997) Identified descending brain neurons control different stridulatory motor patterns in an acridid grasshopper. J Comp Physiol A 180:285–294

Hedwig B, Poulet J (2005) Mechanisms underlying phonotactic steering in the cricket *Gryllus bimaculatus* revealed with a fast trackball system. J Exp Biol 208:915–927

Hedwig B, Poulet JFA (2004) Complex auditory behaviour emerges from simple reactive steering. Nature 430:781–785

Heffner R, Koay G, Heffner H (2007) Sound-localization acuity and its relation to vision in large and small fruit-eating bats: I. Echolocating species, *Phyllostomus hastatus* and *Carollia perspicillata*. Hearing Res 234:1–9

Helversen D, Helversen O (1995) Acoustic pattern recognition and orientation in orthopteran insects: parallel or serial processing? J Comp Physiol A 177:767–774

Hennig RM (1990) Neuronal organisation of the flight motor pattern in the cricket, *Teleogryllus commodus*. J Comp Physiol A 167(5):629–639

Hennig R, Otto D (1996) Distributed control of song pattern generation in crickets revealed by lesions to the thoracic ganglia. Zoology 99:268–276

Hoy R (1978) Acoustic communication in crickets: a model system for the study of feature detection. Fed Proc 37:2316–2323

Huber F (1955) Sitz und Bedeutung nervöser Zentren für Instinkthandlungen beim Männchen von *Gryllus campestris* L. Zeitschrift für Tierpsychologie 12:12–48

Huber F, Thorson J (1985) Cricket auditory communication. Sci Am 253:46–54

Imaizumi K, Pollack GS (2001) Neural representation of sound amplitude by functionally different auditory receptors in crickets. J Acoust Soc Am 109:1247–1260

Kostarakos K, Römer H (2010) Sound transmission and directional hearing in field crickets: neurophysiological studies outdoors. J Comp Physiol A 196:1–13

Kostarakos K, Hedwig B (2012) Calling song recognition in female crickets: temporal tuning of identified brain neurons matches behaviour. J Neurophysiol 32:9601–9612

Kupfermann I, Weiss KR (1978) The command neuron concept. Behav Brain Sci 1:3–10

Kutsch W (1969) Neuromuskuläre Aktivität bei verschiedenen Verhaltensweisen von drei Grillenarten. J Comp Physiol A 63:335–378

Kutsch W, Huber F (1989) Neural basis of song production. In: Huber F, Moore TE, Loher W (eds) Cricket behaviour and neurobiology. Cornell University Press, London, pp 262–309

Larsen O, Kleindienst B, Michelsen A (1989) Biophysical aspects of sound perception. In: Huber F, Moore TE, Loher W (eds) Cricket behaviour and neurobiology. Cornell University Press, London, pp 364–390

Mason AC, Oshinsky ML, Hoy RR (2001) Hyperacute directional hearing in a microscale auditory system. Nature 410:686–690

Michelsen A (1992) Hearing and sound communication in small animals: evolutionary adaptations to the laws of physics. In: Webster DB, Fay RR, An Popper (eds) The evolutionary biology of hearing. Springer, Berlin, pp 61–77

Michelsen A, Löhe G (1995) Tuned directionality in cricket ears. Nature 375:639

Michelsen A, Popov A, Lewis B (1994) Physics of directional hearing in the cricket *Gryllus bimaculatus*. J Comp Physiol A 175:153–164

Montealegre-Z F, Jonsson T, Robert D (2011) Sound radiation and wing mechanics in stridulating field crickets (Orthoptera: Gryllidae). J Exp Biol 214:2105–2117

Murphey R, Zaretsky M (1972) Orientation to calling song by female crickets, *Scapsipedus marginatus* (Gryllidae). J Exp Biol 56:335–352

Nakamura T, Yoshizaki M, Ogawa S, Okamoto H, Shinmyo Y, Bando T, Ohuchi H, Noji S, Mito T (2010) Imaging of transgenic cricket embryos reveals cell movements consistent with a syncytial patterning mechanism. Curr Biol 20:1641–1647

Nocke H (1972) Physiological aspects of sound communication in crickets (*Gryllus campestris* L.). J Comp Physiol A 80:141–162

Nolen T, Hoy R (1984) Initiation of behavior by single neurons: the role of behavioral context. Science 226:992

Olshausen BA, Field DJ (2004) Sparse coding of sensory inputs. Curr Opin Neurobiol 14:481–487

Otte D (1992) Evolution of cricket songs. J Orthop Res 1:25–49

Otto D (1971) Untersuchungen zur zentralnervösen kontrolle der lauterzeugung von grillen. J Comp Physiol A 74:227–271

Otto D (1978) Änderungen von Gesangsparametern bei der Grille (*Gryllus campestris* L.) nach Injektion von Pharmaka ins Gehirn. Verh Dtsch Zool Ges 245

Otto D, Hennig R (1993) Interneurons descending from the cricket subesophageal ganglion control stridulation and ventilation. Naturwissenschaften 80:36–38

Perkel DH, Mulloney B (1974) Motor pattern production in reciprocally inhibitory neurons exhibiting postinhibitory rebound. Science 185:181–183

Petrou G, Webb B (2012) Detailed tracking of body and leg movements of a freely walking female cricket during phonotaxis. J Neurosci Meth 203:56–68

Pollack G (2000) Who, what, where? recognition and localization of acoustic signals by insects. Curr Opin Neurobiol 10:763–767

Pollack G, Faulkes Z (1998) Representation of behaviorally relevant sound frequencies by auditory receptors in the cricket *Teleogryllus oceanicus*. J Exp Biol 201:155–163

Pollack GS (1988) Selective attention in an insect auditory neuron. J Neurosci 8:2635–2639

Pollack GS, Hoy RR (1979) Temporal pattern as a cue for species-specific calling song recognition in crickets. Science 204:429

Popov A, Shuvalov V, Svetlogorskaya I, Markovich A (1974) Acoustic behaviour and auditory system in insects. In: J. Schwartzkopff, (ed) Mechanoreception. Abh Rheinisch-Westfäl Akad Wiss, vol 53. pp 281–306

Poulet JFA, Hedwig B (2002) A corollary discharge maintains auditory sensitivity during sound production. Nature 418:872–876

Poulet JFA, Hedwig B (2006) The cellular basis of a corollary discharge. Science 311:518

Poulet JFA, Hedwig B (2005) Auditory orientation in crickets: pattern recognition controls reactive steering. Proc Nat Acad Sci 102:15665–15669

Reeve RE, Webb BH (2003) New neural circuits for robot phonotaxis. Phil Trans Roy Soc London A 361:2245–2266

Regen J (1913) Ueber die anlockung des weibchens von *Gryllus campestris* L. durch telephonisch übertragene Stridulationslaute des Männchens. Pflüg Archiv Europ J Physiol 155:193–200

Rheinlaender J, Blätgen G (1982) The precision of auditory lateralization in the cricket, Gryllus bimaculatus. Physiol Entomol 7:209–218

Roesel von Rosenhof AJ (1749) Insectenbelustigung zweyter Theil, welcher acht Klassen verschiedener sowohl inlaendischer als auch einiger auslaendischer Insecten enthaelt. Nuernberg, Fleischmann JJ

Rose G, Capranica RR (1983) Temporal selectivity in the central auditory system of the leopard frog. Science 219:1087–1089

Rose GJ, Leary CJ, Edwards CJ (2011) Interval-counting neurons in the anuran auditory midbrain: factors underlying diversity of interval tuning. J Comp Physiol A 197:97–108

Schildberger K (1984) Temporal selectivity of identified auditory neurons in the cricket brain. J Comp Physiol A 155:171–185

Schildberger K, Hörner M (1988) The function of auditory neurons in cricket phonotaxis I. Influence of hyperpolarization of identified neurons on sound localization. J Comp Physiol A 163:621–631

Schildberger K, Huber F, Wohlers D (1989) Central auditory pathway: neuronal correlates of phonotactic behavior. In: Huber F, Moore TE, Loher W (eds) Cricket behaviour and neurobiology. Cornell University Press, London, pp 423–458

Schmitz B, Scharstein H, Wendler G (1982) Phonotaxis in *Gryllus campestris* L. (Orthoptera, Gryllidae). J Comp Physiol A 148:431–444

Schöneich S, Hedwig B (2010) Hyperacute directional hearing and phonotactic steering in the cricket (*Gryllus bimaculatus* deGeer). PLoS ONE 5(12):e15141

Schöneich S, Hedwig B (2011) Neural basis of singing in crickets: central pattern generation in abdominal ganglia. Naturwissenschaften 98(12):1069–1073

Schöneich S, Hedwig B (2012) Cellular basis for singing motor pattern generation in the field cricket (*Gryllus bimaculatus* DeGeer). Brain Behav 2(6):707–725

Sobel EC, Tank DW (1994) In vivo Ca^{2+} dynamics in a cricket auditory neuron: an example of chemical computation. Science 263:823–826

Staudacher E (1998) Distribution and morphology of descending brain neurons in the cricket *Gryllus bimaculatus*. Cell Tiss Res 294:187–202

Staudacher E, Schildberger K (1998) Gating of sensory responses of descending brain neurones during walking in crickets. J Exp Biol 201:559–572

Staudacher EM (2001) Sensory responses of descending brain neurons in the walking cricket, *Gryllus bimaculatus*. J Comp Physiol A 187:1–17

Stumpner A, von Helversen D (2001) Evolution and function of auditory systems in insects. Naturwissenschaften 88:159–170

Stout JF, DeHaan CH, McGhee RW (1983) Attractiveness of the male *Acheta domesticus* calling song to females I. Dependence on each of the calling song features. J Comp Physiol A 153:509–521

Thorson J, Weber T, Huber F (1982) Auditory behavior of the cricket II. Simplicity of calling-song recognition in *Gryllus* and anomalous phonotaxis at abnormal frequencies. J Comp Physiol A 146:361–378

Tschuch G (1977) Der Einfluss synthetischer Gesänge auf die Weibchen von *Gryllus bimaculatus* de Geer (Teil 2). Zool Jb Physiol 81:360–372

Ulagaraj S, Walker TJ (1973) Phonotaxis of crickets in flight: attraction of male and female crickets to male calling songs. Science 182:1278–1279

Weber T, Thorson J (1989) Phonotactic behavior of walking crickets. In: Huber F, Moore TE, Loher W (eds) Cricket behaviour and neurobiology. Cornell University Press, London, pp 310–339

Wendler G, Dambach M, Schmitz B, Scharstein H (1980) Analysis of the acoustic orientation behavior in crickets (*Gryllus campestris* L.). Naturwissenschaften 67:99–101

Wenzel B, Hedwig B (1999) Neurochemical control of cricket stridulation revealed by pharmacological microinjections into the brain. J Exp Biol 202:2203–2216
Wiese K, Eilts-Grimm K (1985) Functional potential of recurrent lateral inhibition in cricket audition. In: Kalmring K, Elsner N (eds) Acoustic and vibrational communication in insects. Parey, Berlin, pp 33–40
Witney AG, Hedwig B (2011) Kinematics of phonotactic steering in the walking cricket *Gryllus bimaculatus* (de Geer). J Exp Biol 214:69–79
Wohlers DW, Huber F (1982) Processing of sound signals by six types of neurons in the prothoracic ganglion of the cricket, *Gryllus campestris* L. J Comp Physiol A 146:161–173
Wyttenbach RA, May ML, Hoy RR (1996) Categorical perception of sound frequency by crickets. Science 273:1542–1544
Zorovic M, Hedwig B (2011) Processing of species-specific auditory patterns in the cricket brain by ascending, local, and descending neurons during standing and walking. J Neurophysiol 105:2181–2194

Chapter 9
Neural Processing in the Bush-Cricket Auditory Pathway

Andreas Stumpner and Manuela Nowotny

Abstract Bush-crickets have their ears in the front tibia with the sensory organ located between two eardrums. An acoustic trachea that amplifies higher frequencies guides sound towards the sensory organ. A linearly arranged set of primary auditory neurons functions like a filter bank, for frequency discrimination and additionally supports intensity discrimination. At the level of attachment cells, a travelling wave might contribute to the frequency tuning of the auditory afferents. The first level of central auditory processing is characterised by convergence of sensory neurons onto thoracic local and intersegmental interneurons and is shaped by presynaptic and postsynaptic inhibition. Interneurons tuned to the carrier frequency are prime candidates for processing intraspecific communication signals. Other neurons are well suited for bat detection and in one neuron auditory stream segregation of conspecific and bat calls was demonstrated. Auditory brain neurons reveal processing properties not encountered in thoracic ganglia.

9.1 Neural Processing in the Bush-Cricket Auditory Pathway

Nearly all species of the Ensiferan taxon Tettigoniidae, also known as katydids or bush-crickets, produce a diverse range of conspecific communication signals (see Chap. 10 by Schul et al.). While few taxa have secondarily lost the capacity to hear

A. Stumpner (✉)
Department of Cellular Neurobiology, Johann-Friedrich-Blumenbach-Institute for Zoology and Anthropology, Georg-August-University Göttingen, Schwann-Schleiden-Forschungszentrum, Julia-Lermontowa-Weg 3, D-37077 Göttingen, Germany
e-mail: astumpn@gwdg.de

M. Nowotny
Department of Cell Biology and Neuroscience, AK Neurobiology and Biosensors, Max-von-Laue-Straße 13, D-60438 Frankfurt am Main, Germany
e-mail: nowotny@bio.uni-frankfurt.de

B. Hedwig (ed.), *Insect Hearing and Acoustic Communication*,
Animal Signals and Communication 1, DOI: 10.1007/978-3-642-40462-7_9,

(Lakes-Harlan et al. 1991), the vast majority possesses broad hearing ranges, which usually extend far into the ultrasonic frequencies (Römer 1987). Production of communication calls and hearing has multiple intraspecific and interspecific functions, like (i) mate attraction which is usually accomplished by singing males, but in some groups females do reply and attract males in return, (ii) male spacing (Bailey and Thiele 1983; Römer and Bailey 1986; Arak et al. 1990), and (iii) aggressive behaviour between males (Morris 1971), as well as (iv) predator avoidance by stopping to sing or evasive flight behaviour (Libersat and Hoy 1991; Schulz and Schul 2001).

9.1.1 Ear Anatomy and Physiology

Tympanic hearing organs of bush-crickets consist of special structures to transmit sound (eardrums and acoustic trachea), and of the peripheral receptor organs in the foreleg tibiae (Fig. 9.1). On each side, an acoustic trachea with several morphological adaptations guides airborne sound towards the peripheral receptor organ. The trachea's opening, the spiracle, is located laterally in the mesothorax posterior to the pronotum and constitutes the main sound input—at least for higher frequencies (Seymour et al. 1978; Heinrich et al. 1993; Michelsen et al. 1994). The affiliated voluminous bullae of the left and right side usually are not in direct contact. In contrast to crickets, a transfer of sound between the two auditory tracheae has not yet been found, but may occur in animals where the large bullae are in direct contact (see Bailey 1990 for a summary). Towards the coxa, the bulla tapers into the acoustical trachea within the foreleg (Fig. 9.2). This particular exponential-horn shape of the auditory trachea amplifies high-frequency sound energy on its way to the receptor organ by 20–25 dB (Fig. 9.2; Lewis et al. 1974; Shen 1993a; Hoffmann and Jatho 1995). The distinct horn shape may not be present in all species of bush-crickets and resonant cavities in the thorax might exist (Nocke 1975; Bailey 1990). In other cases, within the tibia, the acoustic trachea comprises areas without changes of its diameter, where resonant properties may occur (Hill and Oldfiel 1981; Nowotny et al. 2010). Tympanal membranes are located at the anterior and posterior side of each foreleg tibia, with the auditory organ housing the scolopidial mechanoreceptors situated between them. Unlike in many crickets, the two tympanal membranes of bush-crickets are more or less symmetrical and in most species partly hidden under cuticular folds, which leave only a small slit for sound entry. In species with open tympana, the outer surface of the tympanal membranes is subdivided into two areas, a dorsal inner plate (IP) and a surrounding membrane (SM, Fig. 9.1 see also Schumacher 1975). Upon acoustic stimulation, the tympanum moves in the dorso-ventral direction like a hinged flap with the largest displacement in its ventral part (Bangert et al. 1998; Nowotny et al. 2010). The responses of the auditory organ to direct sound input via the tympanal membranes, however, are less sensitive and biased towards lower frequencies (e.g. Lewis 1974, Michelsen and Larsen 1978; Hummel et al. 2011). Recent investigations on *Copiphora gorgonensis* reveal the existence of a special fulcrum, which

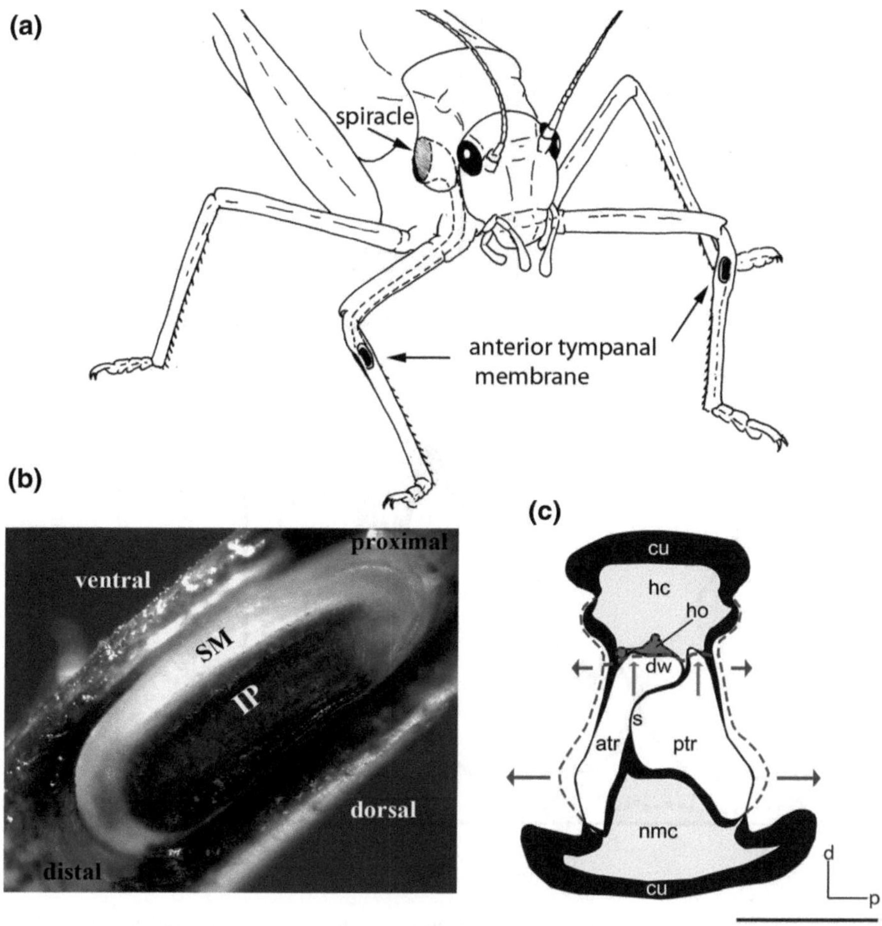

Fig. 9.1 Anatomy of the bush-cricket ear. **a** Lateral view of a bush-cricket, indicating the location of the hearing organ in the foreleg tibia, the sensory neurons are arranged close to the inside of the anterior tympanal membrane, the adjacent acoustical trachea and its opening, the spiracle, are indicated by a stippled line. **b** Photograph of an anterior tympanum. IP = inner plate, SM = surrounding membrane. From Nowotny et al. (2010), with permission. **c** Schematic cross-section of the tibia in the region of the hearing organ. The light grey arrows indicate a possible internal sound pressure acting on the dorsal wall and therefore on the bottom of the hearing organ. Additionally, sound-induced tympanic membrane motion and a possible corresponding stretch-and-pull effect on the dorsal wall as well as the translocation of the hearing organ are indicated (dark grey arrows and dashed line; movement is not drawn to scale). atr = anterior trachea branch, cu = cuticle, d = dorsal, dw = dorsal wall, hc = hemolymph channel, ho = hearing organ, nmc = nerve and muscle channel, p = posterior, ptr = posterior trachea branch, s = septum. Reproduced with permission from Hummel et al. (2011). Scale bar: 0.5 mm

allows a more complex movement of the tympana that ensures an impedance conversion in concordance with an acoustic vesicle, which lies dorsal of the auditory organ and is filled with an oily substance (Montealegre-Z et al. 2012).

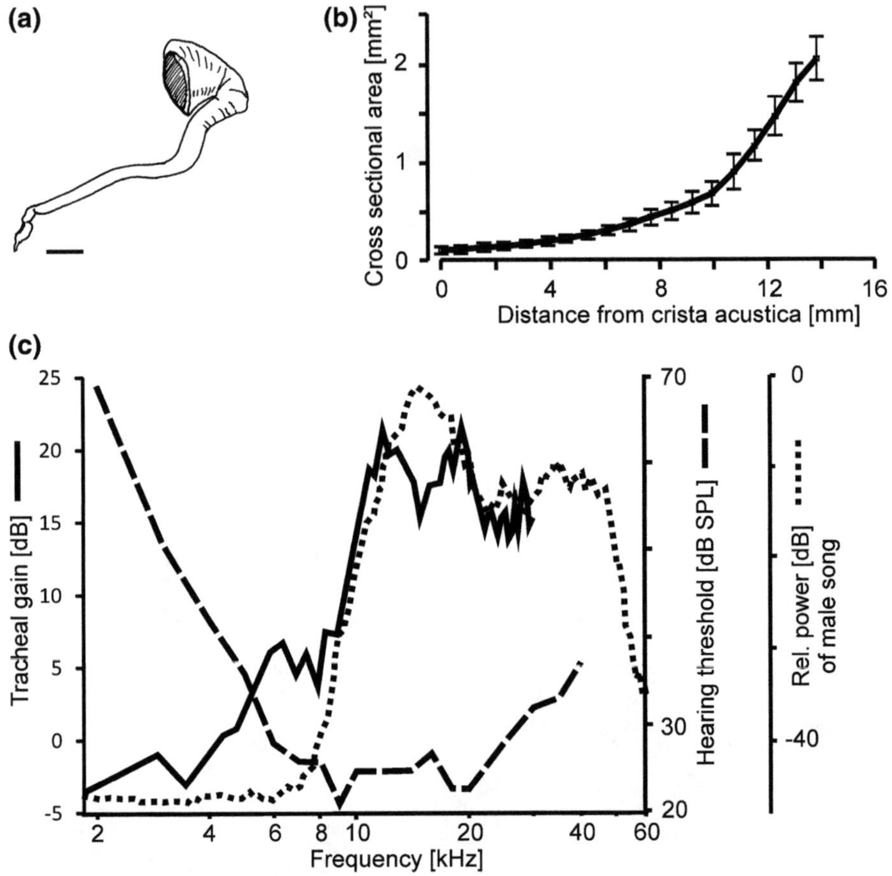

Fig. 9.2 Morphology and function of the acoustic trachea in the bush-cricket *Decticus verrucivorus*. **a** Shape of the trachea from the auditory spiracle to the hearing organ. Bar: 2 mm. **b** Cross-section area of the trachea (mean and SD, n $\geq$ 8). **c** Tracheal gain, measured with a probe microphone, hearing threshold as calculated from the lowest thresholds of recordings from individual sensory neurons, and power spectrum of the male song. Note the different scaling of y-axes. (**a**) Trachea and (**c**) hearing threshold and power spectrum, redrawn with permission from Heinrich et al. 1993; Copyright Acoustical Society of America. (**b**) and (**c**) Tracheal gain, redrawn with permission from Hoffmann and Jatho 1995; Copyright Acoustical Society of America

Bush-crickets comprise complex tibial receptor organs in all three pairs of legs, yet only the foreleg organs are suited to perceive airborne sound with high sensitivity, due to the specific tracheal and tympanal adaptations (Kalmring et al. 1994). Each receptor organ consists, from proximal to distal, of the subgenual organ, the intermediate organ and the crista acustica (Schwabe 1906, Schumacher 1975, 1979; Strauß and Lakes-Harlan 2008a, b, and 2009; see Chap. 2 by Strauß and Lakes-Harlan). Sensory cells of the subgenual organ are linked to the dorsal hypodermis and respond to acceleration induced by substrate-born vibrations up to 5 kHz

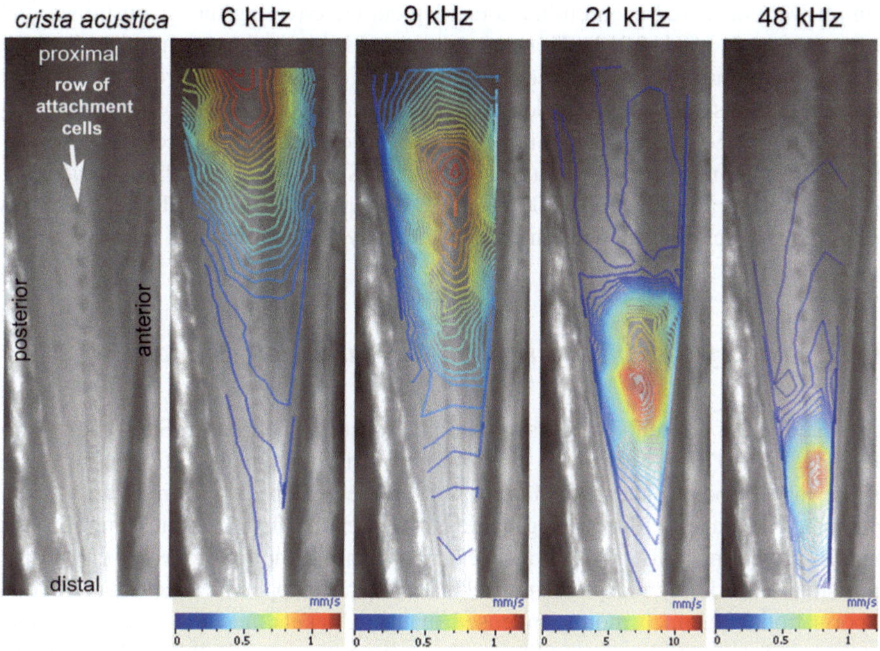

Fig. 9.3 Frequency discrimination along the *crista acustica* by travelling waves on the tectorial membrane in the plane of attachment cells. Pure-tone stimulation leads to travelling waves, exhibiting a tonotopically arranged amplitude distribution along the hearing organ. Coloured lines represent vibration velocity, with *red* for high velocity response and blue for low velocity response (see *colour bars* at the *bottom*). Data are not corrected for air-ringer refractivity

(Kühne 1982; Stumpner 1996). The subsequent intermediate organ seems to merge with the distally adjacent crista acustica; their receptors are distinguished based on their spatial arrangement and physiology (Field and Matheson 1998; Stölting and Stumpner 1998). In the crista acustica, approximately 20–60 linearly arranged primary neurons provide a filter bank covering a broad frequency spectrum from about 5 to 100 kHz (Römer 1987; Römer et al. 1988; Lin et al. 1993). The receptors of the crista acustica are tonotopically organised, with their best frequencies increasing from proximal to distal. This was shown by electrophysiological recordings from the soma (e.g. Oldfield 1982) or from the terminal arborisations in the thorax in conjunction with labelling the peripheral soma position (Stumpner 1996; Stölting and Stumpner 1998). The tuning of the sensory neurons is accomplished via sound-induced travelling waves along the crista acustica (Fig. 9.3; Palghat Udayashankar et al. 2012), leading to place-specific amplitude responses in dependence of frequency. With an increase in frequency the travelling waves are characterised by an increase in velocity and a decrease in wavelength. These results coincide with morphological investigations (e.g. Rössler et al. 1994) assigning to structural features of the crista acustica the largest importance for frequency tuning. In the study on *C. gorgonensis* a travelling wave and tonotopic representation of the applied pure-tone stimuli were observed even on the dorsal leg cuticle covering the "acoustic vesicle" (Montealegre-Z et al. 2012). However, some evidence indicates that also intrinsic factors may contribute to the frequency tuning of the sensory neurons, as adjacent sensory neurons may exhibit identical tuning but clear differences in sensitivity (e.g. Oldfield 1984) and their tuning may be little affected by mechanical disruption of the tympanic membrane (Oldfield 1985; Kalmring et al. 1993). In contrast, Korsunovskaya and Zhantiev (2007) report a temperature independence of sensory cell tuning, which implies that intrinsic factors may not be important for tuning.

9.1.2 Sensory Cells: Structural and Functional Tonotopy

The auditory afferents are scolopidial sensory neurons (Schwabe 1906; Schumacher 1979) as found in chordotonal sense organs in different locations of the insect body (Field and Matheson 1998; see Chap. 2 by Strauß and Lakes-Harlan). Their axons join the leg nerve, and enter the first thoracic ganglion as a bundle projecting into the anterior ventral neuropile, homologous to the projection sites of auditory receptors in crickets and grasshoppers. However, the axonal arborisations are much more condensed than in crickets, and the whole neuropile appears to be rotated by about 60° anteriorly (Imaizumi and Pollack 2005). The filter bank-like organisation of the sensory neurons in the auditory organ with specific frequency tuning of the sensory neurons is matched and preserved by a tonotopic projection pattern of their axons (Zhantiev and Korsunovskaya 1978; Oldfield 1983; Römer 1983). Whether each sensory neuron responds most sensitive to a slightly different carrier frequency and covers a slightly shifted projection area as compared to the neighbouring cell (Stölting and Stumpner 1998), or whether at least some adjacent neurons may have identical tuning and a similar projection site is controversial and might differ between species (Oldfield 1983; Lin et al. 1993; Stölting and Stumpner 1998). Interindividual variability, however, hampers such analyses (Ostrowski and Stumpner 2010) and makes conclusions of purely physiological studies less reliable. A sensory organ representing a filter bank might provide frequency analysis with high precision in the first place (Pollack and Imaizumi 1999). However, changing frequency tuning over a set of neurons can also serve as mechanism for better intensity coding (Hardt 1988; Hennig et al. 2004; Höbel and Schul 2007). This is demonstrated with the intensity response functions of nine individually identified sensory neurons of the bush-cricket *Ancistrura nigrovittata* (Phaneropterinae) covering with their dynamic responses a range of more than 60 dB SPL when stimulated at one carrier frequency (Fig. 9.4). In basal Prophalangopsidae (formerly called Haglidae; e.g. genus *Cyphoderris*) the whole ear is tuned to low frequencies and differential tuning between adjacent cells is limited (Mason 1991; Mason et al. 1999; Mason and Faure 2004).

Auditory afferents exhibit tonic spike responses with an individual dynamic range of 15–25 dB (Rheinlaender 1975; Oldfield 1983) and with clear adaptation (Ostrowski 2006). The highest spike frequencies range between 300 and 400 Hz, independent of

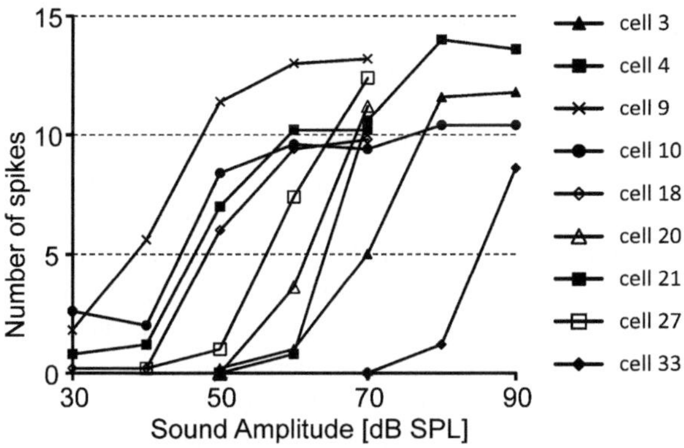

Fig. 9.4 Intensity response functions of nine individually identified afferent neurons (out of ca. 36) in female *Ancistrura nigrovittata* (Phaneropterinae). The cell number indicates the position in the crista acustica from proximal to distal. Stimuli were pure-tone pulses (16 kHz, 50 ms). The dynamic range of these sensory neurons together covers more than 60 dB

the position of the neuron within the hearing organ and of the best frequency of the neuron. Adaptation for longer lasting stimuli reduces the response by about 30–40 %; steady state is reached within 50–100 ms of a sound stimulus (the louder the stimulus the later the steady state is reached). The average adaptation time constant for stimuli at least 15 dB above threshold is ~35 ms (Jan Benda; Tim Ostrowski; Andreas Stumpner, unpublished data; Benda and Herz 2003). Adaptation seems to be independent of carrier frequency or position of the cell in the crista acustica (Ostrowski 2006).

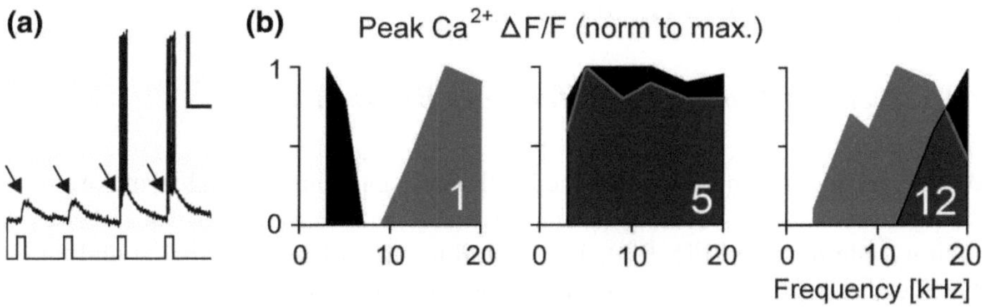

Fig. 9.5 Primary afferent depolarisation (PAD) of sensory neurons. **a** Intracellular recording close to the axon terminal in the prothoracic ganglion of a sensory neuron with the cell body located at position 21 of 24 in the crista acustica of *Pholidoptera griseoaptera* (Decticinae). Stimuli: 50 kHz, 60 and 70 dB SPL, two stimuli each. The arrows point at graded potentials. Bars: 25 mv and 100 ms. From Stölting 1996, with permission. **b** Ca^{2+}-signals revealing frequency tuning of afferents of *Mecopoda elongata* (Phaneropterinae) in the intact system (*black*) and frequency tuning of the PADs (*grey*) determined after removal of the ipsilateral ear. Responses of three different auditory afferents (out of twelve tested) are arranged according to their tuning in the intact system. Measurements are the averaged Ca^{2+} signals of 3 sound presentations. From Baden and Hedwig (2010) with permission

Graded depolarizing potentials in terminals of sensory cells have been noted in several bush-cricket species (e.g. *Pholidoptera griseoaptera*, Stölting 1996, see Fig. 9.5) and GABA-ergic interneurons that likely mediate the input of the graded potentials have been found to form synapses with sensory neurons in *Tettigonia cantans* (Hardt and Watson 1999). These graded potentials have been analysed in some detail by Baden and Hedwig (2010) in the bush-cricket *Mecopoda elongata* (Fig. 9.5). The potentials were identified as typical primary afferent depolarisations (PAD, Watson 2002) with the effect of lowering the action potential amplitude by approximately 10 % when recorded next to the axon terminals. PADs evoked by sensory input from both ears, are sensitive to picrotoxin, a general blocker of chloride channels in insects (Burrows and Laurent 1993; Lynch et al. 1995), and are not linked to spiking activity of the targeted afferent. Tuning of excitation and presynaptic inhibition—as judged from experiments with Ca^{2+}-imaging—coincided largely in some sensory neurons most sensitive to medium frequencies, but differed drastically in other neurons, which were activated by frequencies below 10 kHz and received PADs by frequencies above 10 kHz (Fig. 9.5). In general, it seemed that low-frequency neurons received high-frequency PADs and vice versa. Moreover, the relatively wide branching mid-frequency sensory neurons experienced different strength of PAD on different terminals—more high-frequency tuned PADs in the anterior ("low frequency") portion of the neuropile and more low-frequency tuned PADs in the posterior portion of the neuropile (Baden and Hedwig 2010). These PADs may sharpen frequency tuning by reducing the effect of synaptic transmission at one flank of auditory tuning, but they may also function as a gain control mechanism—especially in those mid frequency neurons that are tuned to the conspecific song.

9.2 First Order Interneurons

9.2.1 Neuron Types and Differences Between Taxa

Within the auditory neuropile of the prothoracic ganglion, axonal terminations of the auditory afferents overlap at least partially with dendrites of four classes of auditory interneurons (Fig. 9.6). The overall number of neuron types found within these four classes most likely differs between groups of bush-crickets.

(i) **Local neurons** with at least four different types: omega neuron (ON; Zhantiev and Korsunovskaya 1983), other bilateral neurons (SN1; Zhantiev and Korsunovskaya 1990), hemisegmental neurons (SN2; Stumpner 1995) and DUM-cells (Stritih and Stumpner 2009). ON was found in each species of bush-crickets, in which single cell recordings in the prothoracic ganglion were performed.
(ii) **Ascending neurons**, with a prothoracic soma and an axon ascending towards the brain. There may be at least four to five ANs in some groups or species (Rheinlaender and Kalmring 1973; Shen 1993b; Stumpner and Molina 2006). At least two different types AN1 and AN2 have been found, whenever there

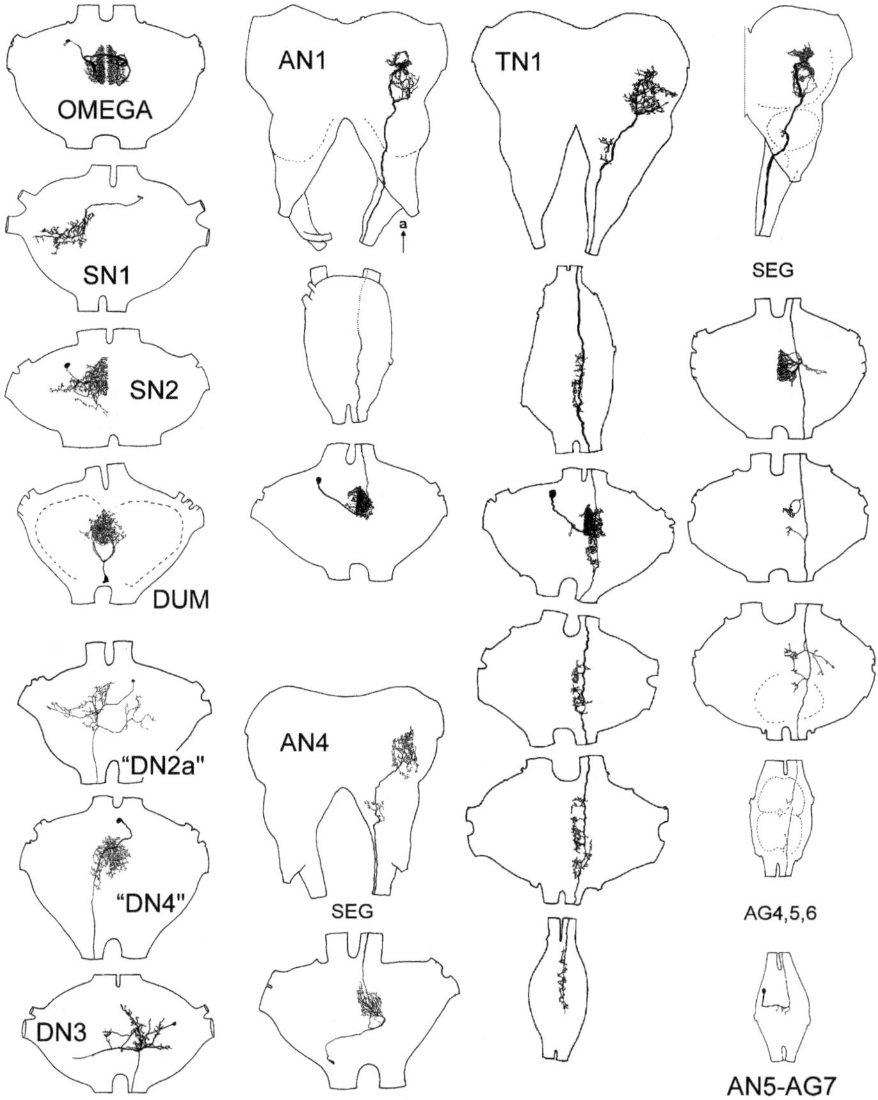

Fig. 9.6 Morphology of different auditory interneurons in Phaneropterinae. First column, prothoracic ganglion, upper four: local neurons. Omega from Stumpner (2002); SN1 from Zhantiev and Korsunovskaya (1990); DUM from Stritih and Stumpner (2009); with permission. First column lower three: Descending neurons; the names of DN2 and DN4 are preliminary. DN3 from Zhantiev and Korsunovskaya (1990), with permission. Second row: Ascending neurons; AN1 from Stumpner and Molina (2006), with permission. Third and fourth column: TN1 and AN5-AG7 from Stumpner (1999), with permission. Abbreviations: a = anterior; AG: abdominal ganglia; AN = ascending neuron; DN = descending neuron; DUM = dorsal unpaired median; SN = segmental neuron; TN = T-shaped neuron, SEG: subesophageal ganglion, T: thoracic ganglion; numbers refer to the respective segment

was an extensive study of prothoracic neurons. These may be homologous to AN1 and AN2 of crickets (Stumpner 1997).

(iii) **Descending neurons**, with an axon descending towards the posterior thoracic and abdominal ganglia. In the prothoracic ganglion, they seem to form a cluster with an anterio-medial soma position (Rheinlaender and Kalmring 1973; Römer 1987; Sickmann 1997), but some may have a soma in a more lateral position (Zhantiev and Korsunovskaya 1990).

(iv) **T-fibre neurons**, with both an ascending and a descending axon, typically, but not always with the soma in the prothoracic ganglion. At least one type, the T-shaped neuron 1 (TN1) has been described for several species. Often it is just identified by its large axon leading to easily detectable spikes in extracellular recordings from the neck connective (Suga and Katsuki 1961a). From studies without morphological identification, the existence of up to eight T-shaped neurons has been proposed for Decticinae (Rheinlaender and Kalmring 1973). On the other hand, considerable interindividual variability of TN1 has been found (Stumpner and Molina 2006; Johannes Schul, personal communication).

So far only few species have been studied providing a combination of morphological and physiological data for auditory interneurons. Therefore, some apparent differences in the number of neuron types between groups of bush-crickets may just represent a lack of knowledge. There might be a tendency, though, that species with more complex songs have more neurons. Nevertheless, compared to other insect species like grasshoppers (see Chap. 11 by Ronacher) the number of auditory neurons seems to be considerably lower.

In species, where a larger number of neuron types have been physiologically and morphologically identified (e.g. in *A. nigrovittata*, but see also Shen 1993b),

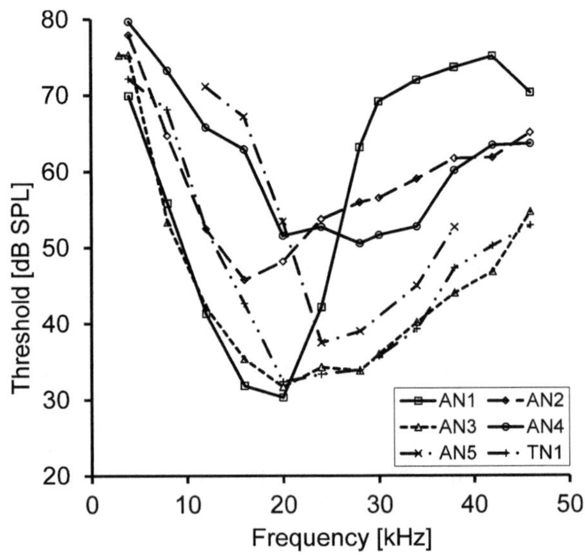

Fig. 9.7 Frequency tuning curves of six different neurons with axons ascending from the prothoracic ganglion to the brain in *A. nigrovittata*. Stimuli were pure-tone pulses (50 ms). Modified with permission after Stumpner and Molina (2006)

each of these types represents a unique combination of carrier frequency tuning, sensitivity and temporal response properties (Fig. 9.7). The neurons cannot be simply divided into high-frequency and low-frequency channels, but represent (i) broadband neurons with tonic response pattern (ON1, SN2, certain DN, AN3), (ii) very low-frequency neurons, mostly also responsive to vibration (certain DN), (iii) low-frequency neurons with differences in sensitivity and temporal pattern processing (certain DUM, AN1; AN2 in some species), (iv) phasic to (phasic-)tonic broader tuned high-frequency neurons (certain DUM, AN2 in some species, AN4, TN1, certain DN) and (v) phasic narrower tuned high-frequency neurons (AN5-AG7). For similar results from studies without morphological identification see, e.g. Oldfield and Hill (1983) for *Caedicia simplex* or Rheinlaender and Kalmring (1973) for *Decticus verrucivorus.* Of course these categories are artificial and the discrimination between more "phasic" and more "tonic" responses may depend on carrier frequency and intensity. Nevertheless, in comparison to auditory afferents several new properties arise on the first level of auditory processing.

9.2.2 Directionality

Nearly all prothoracic neurons, with exception of DUM-neurons, exhibit a clear preference for stimuli from the side, where the dendrites with the major excitatory input sites are housed. This preference is stressed by contralateral inhibition, which has been demonstrated for ON, AN1 and TN1 (Suga and Katsuki 1961a; Rheinlaender et al. 1972; Stumpner and Molina 2006). Directional processing has most convincingly been studied in the ON neuron, which is the most prominent and best characterised local auditory neuron (Zhantiev and Korsunovskaya 1983; Römer 1985). It has its name from its typical shape and from its presumed homology to the omega neuron 1 (ON1) of crickets (Popov et al. 1978; see Molina and Stumpner (2005) for a survey of the literature). It shows similarities also to the cricket ON1's physiology by responding tonic to stimuli over a wide range of frequencies and by showing a strong directional sensitivity. When stimulated via the soma-contralateral ear alone, it typically exhibits clear inhibitory postsynaptic potentials (IPSPs) in dendritic recordings (Schul 1997). These IPSPs have been demonstrated to originate from the mirror image omega neuron (Römer and Krusch 2000; Molina and Stumpner 2005), as has been demonstrated in crickets before (Selverston et al. 1985). The recurrent inhibition between the omega neurons leads to a reliable coding of small bilateral intensity differences of ~1 dB (Stradner and Römer 2008), but also to a clear division of the auditory scene around the bush-cricket into a left and a right hemisphere (Römer and Krusch 2000; compare with Pollack 1988 for crickets). Additionally, the inhibition helps to reduce interference in the coding of signals arriving at the specimen from different directions. An important role of contralateral inhibition in ON was discussed in the context of leader preferences, where the follower signal was mostly suppressed in ON-responses (Römer et al. 2002; Siegert et al. 2011).

These directional mechanisms are complemented by another inhibitory effect with a time constant in the range of seconds gradually leading to a clear preference for the louder of two signals (Römer 1993; Römer and Krusch 2000). The mechanism has been presumed to be similar to the one described for the cricket *Acheta domesticus* (Sobel and Tank 1994). Here, with stronger stimulation a rise in intracellular Ca^{2+}-concentration leads to a Ca^{2+}-dependent increase of K^{+}-conductance and thereby to a gradual hyperpolarisation of the neuron.

Like in crickets, the AN1 is inhibited by the contralateral ON (Molina and Stumpner 2005; for crickets Horseman and Huber 1994; Stumpner et al. 1995). This inhibition increases the directionality of AN1, but may also serve as kind of gain control—at least for neurons with a saturating intensity response function—(Reeve and Webb 2003) as based on more hypothetical technical considerations.

9.2.3 Processing of Carrier Frequency and Temporal Patterns

9.2.3.1 Carrier Frequency Processing

For some thoracic neurons, the extent of dendrites in the auditory neuropile allows a prediction of their carrier frequency tuning due to their overlap with the tonotopically organised axonal arborisations of the sensory cells (Römer 1987; Stumpner 1997). This may be more clearly expressed in species of the taxon Tettigoniinae (Römer et al. 1988), than in Phaneropterinae (Stumpner and Molina 2006). Interneurons do not only receive input from the auditory afferents but most likely from other interneurons as well. Correspondingly dendrites extend through nearly the whole neuropile also in neurons, which are only sensitive to a restricted range of frequencies like the TN1, AN2 and AN4 neurons in *A. nigrovittata* (Stumpner 1999; Stumpner and Molina 2006). While it is not clear, whether such interneuronal input is excitatory at least in some neurons, it is obvious, especially from dendritic recordings, that there is prominent sound frequency-dependent inhibition in the majority of neurons (Römer 1987; Schul 1997; Stumpner 1997, 1998; Triblehorn and Schul 2009). This inhibition sometimes leads to nonlinear intensity response functions with pronounced optimum-type curves, as described in early extracellular recordings of unidentified neurons (Rheinlaender 1975). It was demonstrated for a group of closely related Phaneropterinae (genera Ancistrura, Barbitistes and Leptophyes) that the AN1 neuron—like in crickets—is quite precisely tuned to the carrier frequency of the male song and that frequency-dependent inhibition is an important part of this species-specific tuning (Hardt 1988; Stumpner 2002; for crickets, e.g. Schildberger 1984; Kostarakos et al. 2009).

The effect of frequency-dependent inhibition of interneurons has been directly demonstrated for some bush-cricket species. Frequency tuning of auditory neurons may broaden considerably after application of picrotoxin a pharmacological blocker of chloride channels (Suga and Katsuki 1961b; Stumpner 1998) and species-specific differences may be completely lost (Stumpner 2002). In the AN1 neuron of *A. nigrovittata* a tuning based on excitation from about 9–10 sensory neurons is reduced by this frequency-dependent inhibition to a tuning representing that of one or a few sensory neurons only (Ostrowski and Stumpner 2010). On the one hand, it may be advantageous to drive excitation by a larger number of cells, probably to reduce noise (Machens et al. 2001; see Chap. 11 by Ronacher), and then let inhibition sharpen the tuning. On the other hand, it might be easier during evolution to change an inhibitory input potentially evoked by a single interneuron only than to change excitation by a number of sensory cells. In one species of Phaneropterinae (*Caedicia simplex*) an intersegmental interneuron is even sharper tuned than a single sensory neuron (Oldfield and Hill 1983); also this interneuron receives strong inhibition as exemplified, e.g. by two-tone inhibition. Among the at least six neurons with an axon ascending to the brain in Phaneropterinae, five exhibit clear indication of frequency-dependent inhibition, thereby stressing the importance of this mechanism. Frequency-dependent inhibition of interneurons is also described in grasshoppers (Römer et al. 1981) and in crickets (Schildberger 1984; Stumpner et al. 1995).

Intensity response functions of one neuron type measured at different frequencies often differ clearly in shape, which is a direct consequence of frequency-dependent inhibition (Rheinlaender and Kalmring 1973; Oldfield and Hill 1983; Stumpner and Molina 2006). This inhibition may have the effect that the respective neuron (like AN1 of *A. nigrovittata*, Stumpner 1997) stops firing at high intensities of a conspecific call, which will occur when a female approached a conspecific singer to a low distance. However, at close distances vibratory components of a song should become detectable, and these have been reported to improve the responses of auditory interneurons in certain bush-crickets (Kalmring and Kühne 1980). Optimum-type response functions can also contribute to determining distance of a conspecific neighbour. As broadband signals change their spectral content with distance due to the excess attenuation of high frequencies, the concomitant inhibitory influence on a neuron's responses and therefore its activation changes as well. The activity of a set of frequency-dependent neurons can then represent the distance of the singer (Römer 1987 for *Mygalopsis marki*).

Which local neurons are the source of such inhibition? Local auditory neurons, other than the Omega neuron, have been described for some Phaneropterinae, but

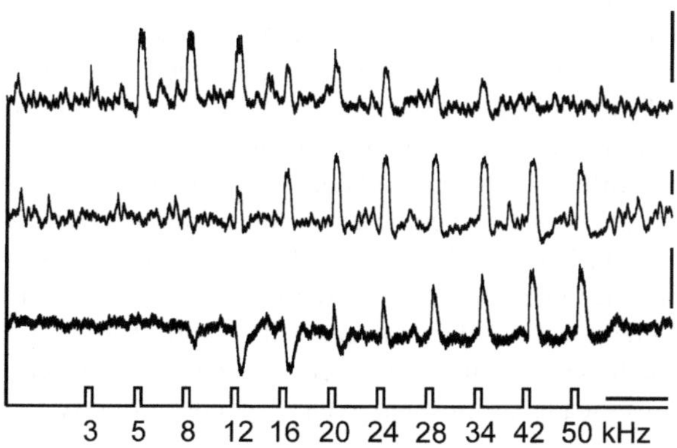

Fig. 9.8 Graded potentials recorded in the soma of three different DUM-type neurons of one individual *A. nigrovittata* male in response to 70 dB SPL stimuli at the frequencies indicated. Note the differences in tuning and the occurrence of clear IPSPs in the lowest trace. Bars: 10 mV and 500 ms

were not studied systematically (Zhantiev and Korsunovskaya 1990; Stumpner 1995; Stritih and Stumpner 2009). Among these are dorsal unpaired medium neurons, i.e. DUM-type neurons (Hoyle et al. 1974; Bräunig and Pflüger 2001), which may belong to the group of GABAergic DUM-cells described for other Orthoptera (Thompson and Siegler 1991). In *A. nigrovittata*, more than 10 such neurons seem to form a cluster, the members of which may differ in carrier frequency tuning (Fig. 9.8). So far, they are the prime candidates for frequency-specific inhibition at low and high frequencies. IPSPs, recorded at one dendritic position in AN1 of *A. nigrovittata*, for example, differ clearly in shape at low and high frequencies indicating different (DUM?-) neurons evoking these inhibitions (Stumpner 1997).

9.2.3.2 Temporal Processing

Relatively few data are available for the properties of interneurons in the domain of temporal processing. The two identified ascending neurons of Tettigoniinae AN1 and AN2 appear to respond tonic without obvious filtering properties (Schul 1997). In contrast, extracellular recordings in Decticinae and intracellular recordings in Phaneropterinae indicated ascending neurons with tonic to phasic-tonic or phasic responses (Rheinlaender and Kalmring 1973; Oldfield and Hill 1983; Stumpner and Molina 2006). One has to be careful, though, with such general classifications, since the temporal response properties strongly depend on carrier frequency and sound intensity (Rheinlaender and Kalmring 1973; Rheinlaender 1975; Libersat and Hoy 1991; Stumpner 1997). A direct correspondence between the species-specific song pattern and temporal processing of neurons at the thoracic level appears to be much less obvious than for carrier frequency processing, even though some neurons' activity reflects the temporal pattern of male songs quite well (Kalmring et al. 1997 for the Phaneropterine *Polysarcus denticauda*, Schul (1997) for the Tettigoniine *T. viridissima*, Stumpner and Molina (2006) for the Phaneropterine *A. nigrovittata*). In various species of Neoconocephalus, which in their songs produce pulse rates of more than 100 Hz (see Chap. 10 by Schul et al.), the response properties of AN1 correlate to some degree with the conspecific song patterns and behavioural preferences (e.g. Triblehorn and Schul 2009). It was reported that spontaneous activity recorded in thoracic connectives from rhythmically active neurons, which do not directly respond to sound, changes highly specifically in response to conspecific songs (Zhantiev et al. 2004). What the potential source and function of such changes is, remains to be elucidated.

Although no neuronal data are available, a spectacular case of song recognition should be mentioned here. An Australian bush-cricket of the taxon Listroscelidinae (*Chlorobalius viridis*) responds to male cicada songs and places a short reply click within the complex song, with exactly the same timing as female cicadas of the same species (Marshall and Hill 2009). The goal of this behaviour is to attract, catch and devour the male cicadas. Chlorobalius can do this successfully with various cicada species and even responds correctly to songs of species from outside

Australia, which they or their ancestors could never have heard before. They fail only occasionally in very complex songs. This behaviour indicates the potential to recognise complex songs even of a different insect taxon.

9.2.3.3 Interplay of Carrier Frequency and Amplitude Modulations in TN1

Special attention has been given to the T-fibre (TN1, Fig. 9.6), which can be recorded extracellularly in the neck connective (Suga and Katsuki 1961a, b). It is typically most sensitive in the near ultrasound, may have thresholds at 50–60 dB SPL up to 100 kHz and has shortest latencies of 11–12 ms in the ultrasonic range when measured in the neck connective. Due to these characteristics, the TN1 has been proposed to be the prime candidate for bat-avoidance behaviour (Libersat and Hoy 1991; Faure and Hoy 2000a, b; Schul et al. 2000; ter Hofstede et al. 2010). This is strongly supported by its responses, which are typically more pronounced at ultrasonic frequencies. TN1 follows pulse repetition rates up to 30 Hz with at least one spike per pulse. Short pulses with fast rise and fall times release stronger responses than other stimuli and integration time constants are around 6 ms in the ultrasound, but around 15 ms at a sonic frequency (Faure and Hoy 2000a). These data fit nicely to TN1 recordings in Neoconocephalus (Schul and Sheridan 2006) demonstrating that the neuron has properties, which have been described as auditory scene segregation in psychoacoustic studies of vertebrates. In the majority of species, TN1 will not only respond to bat sounds, but also to conspecific songs due to its relatively broad tuning and also due to the fact that bush-crickets often produce songs with higher sonic or ultrasonic components (Heller 1988). The song of Neoconocephalus is in the audio range and comprises relatively high pulse rates—much higher than occurring in normal bat calls. TN1 strongly habituates to such high pulse rates and therefore, when bat calls and songs are presented together, the neuron selectively responds to the bat calls only. This is even true, when the carrier frequencies of bat calls and song are reversed stressing the importance of the temporal pattern. The TN1 selective response only breaks down when both stimuli are presented at the same carrier frequency. Regionalised adaptation processes on the dendrites of TN1 might play a decisive role in this scene analysis (Schul et al. 2012). Are there species-specific differences in TN1 response properties, which might correlate with behavioural differences in the context of singing and bat avoidance? Actually, high TN1 spike rates in response to bat calls have been found in species, which stop singing, when hearing bat cries. However, high spike rates in response to bat calls occur also in some other bush-crickets without such behaviour, indicating that TN1 may show specific adaptations for predator avoidance, but likely is not a neuron solely functioning as "bat detector" in all species (ter Hofstede and Fullard 2008; ter Hofstede et al. 2010), It actually may as well be seen as a "novelty-neuron" detecting specific changes in the auditory scene (Schul et al. 2012).

9.3 Brain Neurons: Types, Frequency Processing and Temporal Processing

Behavioural data indicate that the brain is an important station of auditory processing in insects (Bauer and von Helversen 1987; Pires and Hoy 1992). Processing of auditory information in the brain has been studied in crickets (see Chap. 8 by Hedwig), little in grasshoppers (Boyan et al. 1993) and in bush-crickets (Ostrowski 2009; Ostrowski and Stumpner 2010). In the brain axonal arborisations of some ascending auditory bush-cricket neurons receive inputs. AN2 of *A. nigrovittata* (Phaneropterinae), for example, has a distinct area of branches with fine, presumably postsynaptic endings in the deutocerebrum dorsal to the antennal lobes (Stumpner and Molina 2006). Casual tests did not reveal, which information may be processed in this area. In the axon of the TN1-neuron of *A. nigrovittata* graded depolarisations of substantial size have been recorded in the brain (Ostrowski 2009). The effect of these potentials is an increase in peak values of action potentials when riding on the graded potential. This indicates passive propagation in these branches. Due to its time course, the graded potential might be involved in strengthening the detection of females reply clicks, to which TN1

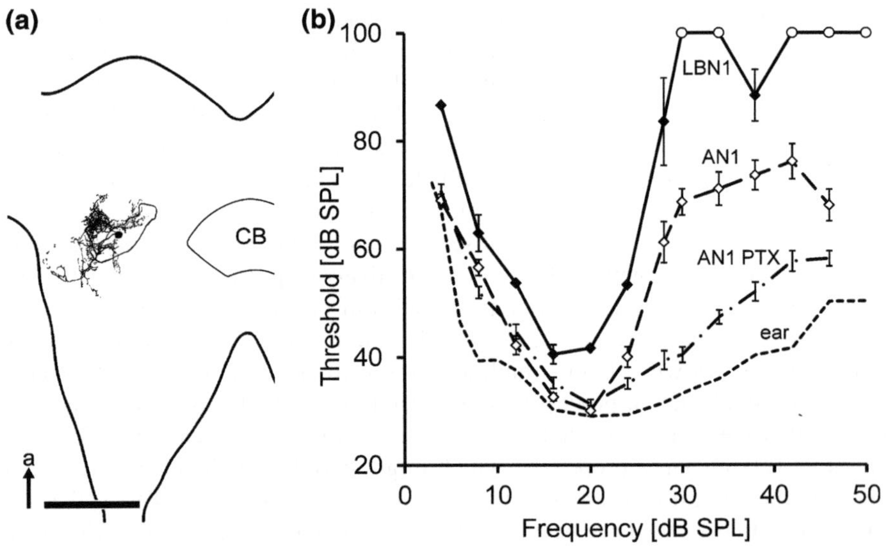

Fig. 9.9 **a** Morphology of the LBN1-neuron in the brain of *A. nigrovittata*. a = anterior; CB = central body; bar = 250 μm. **b** Frequency tuning of LBN1, the ascending neuron AN1 in the intact situation and following application of picrotoxin (PTX), which eliminates frequency-specific inhibition, and of the whole ear as derived from the lowest threshold of 86 sensory cells, recorded in males and females. Error bars show standard error of the mean. Data points with open circles in LBN1 indicate that no response was elicited between 30 and 90 dB SPL at that frequency. From Ostrowski and Stumpner (2010); with permission

responds very well. Female reply clicks, are triggered by a specific element in the male song and occur in a specific time window only (see, e.g. Heller and von Helversen 1986; Zimmermann et al. 1989).

One recent study on sound-activated neurons in the brain of *A. nigrovittata* (Ostrowski 2009) revealed at least ten different types, which surely do not represent the complete set of neurons. Activity of several of these brain neurons reveals features not encountered at the thoracic level. With respect to frequency processing, a specialised neuron (LBN1, Ostrowski and Stumpner 2010; Fig. 9.9) has been identified and physiological and morphological evidence indicates it being directly postsynaptic to the AN1-neuron. AN1 in female *A. nigrovittata* shows the best carrier-frequency tuning for male song on the thoracic level. Sharpening of frequency tuning occurs already on the first level of processing in the thorax, but LBN1 shows additional sharpening of its frequency tuning. First, spike frequency in AN1 has to exceed about 270 Hz to reliably elicit one spike in LBN1. As a consequence, the threshold of LBN1 is approximately 10 dB higher than the threshold of AN1 and ultrasonic frequencies, which elicit only weak responses in AN1, are nearly not responded to by LBN1. Second, frequency-dependent IPSPs occur in LBN1 as in AN1. An effect of blocking this inhibition like in AN1 has not been demonstrated yet for LBN1. Nevertheless, this indicates redundant mechanisms of frequency processing on consecutive levels in the auditory pathway. The carrier-frequency tuning of LBN1 fits quite well to behavioural tuning in this species (Fig. 9.9, Dobler et al. 1994b).

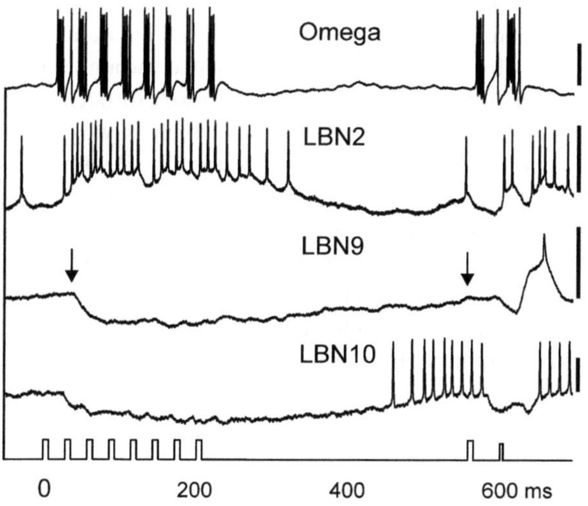

Fig. 9.10 The response of the local thoracic ON neuron and of local brain neurons LBN2, LBN9, LBN10 from *A. nigrovittata* to an artificial duet between a male (16 kHz, first 9 pulses) and a female (28 kHz, short pulse at 600 ms). Note the long lasting depolarisations and hyperpolarisations in the brain neurons. The trace shown for LBN9 is the average of five responses—the arrows indicate beginning and end of the long lasting IPSP. Bars: 25 mV. LBN from Ostrowski (2009) with permission

In addition to brain neurons that are tuned to the 16 kHz carrier frequency of the male song of *A. nigrovittata* and those which are tuned to the 28 kHz female reply click, a neuron was described tuned to both carrier frequencies. It was not just broadly tuned like certain ascending neuron, but intermediate frequencies between 16 and 28 kHz elicited a lower response. Such brain neurons might be suited to code for a duet between the sexes (Fig. 9.10). As for temporal patterns, long-time constants have been seen, e.g. in the slow decrease of inhibition by a male song (Fig. 9.10). Such long lasting processes correspond to the 350–400 ms long interval in the male song between a chirp-like element and a trigger pulse that elicits the female reply click. This 400 ms-interval obviously is represented in the female recognition system, since motivated females may produce reply clicks about 400 ms after a male chirp even when the male trigger pulse is missing (Dobler et al. 1994a). Another local brain neuron exhibited an optimum-type response to varied pulse duration within a chirp. This closely corresponds to the female recognition behaviour, while in neurons at the thoracic level such a correspondence has not been found (Stumpner and Molina 2006).

Certain acoustic behaviours in bush-crickets show extremely short latencies—especially the response clicks of females, which fall into a species-specific time window following the male song (see above, Heller and von Helversen 1986; Zimmermann et al. 1989). In *A. nigrovittata* the shortest replies of females in response to a male trigger signal coincide with the earliest action potentials in the brain neuron LBN1 elicited by the same stimulus. Therefore, one has to conclude that activity of LBN1 cannot directly cause the behavioural response in females, even though LBN1 properties exhibit the closest correspondence to song recognition. An explanation might be that triggering of the female-response clicks is not directly controlled by brain neurons, but rather gated through a more general command from the brain, which allows thoracic reflexes to occur. This requires that males do not just signal once, but repeatedly, which is very common. A similar mechanism has been proposed for cricket phonotaxis (Poulet and Hedwig 2005). Although some auditory neurons forward activity from the prothoracic to the mesothoracic ganglion (e.g. TN and DN neurons), response properties directly corresponding to the observed behaviours have not been encountered yet. This may represent a principal limitation of such intracellular studies—as the tethered bush-cricket is not in the state of searching for a mating partner appropriate commands from the brain may be inhibited.

References

Arak A, Eiriksson T, Radesäter T (1990) The adaptive significance of acoustic spacing in male bushcrickets *Tettigonia viridissima*: a perturbation experiment. Behav Ecol Sociobiol 26:1–7

Baden T, Hedwig B (2010) Primary afferent depolarization and frequency processing in auditory afferents. J Neurosci 30:14862–14869

Bailey WJ (1990) The ear of the bushcricket. In: Bailey WJ, Rentz DCF (eds), The Tettigoniidae. Biology, Systematics and evolution. Crawford House Press, Bathurst, pp 217–247

Bailey WJ, Thiele DR (1983) Male spacing behavior in the Tettigoniidae: an experimental approach. In: Gwynne DT, Morris GK (eds) Orthopteran mating systems: sexual competition in a diverse group of insects. Westview Press, Colorado, pp 163–184

Bangert M, Kalmring K, Sickmann T, Stephen R, Jatho M, Lakes-Harlan R (1998) Stimulus transmission in the auditory receptor organs of the foreleg of bushcrickets (Tettigoniidae) I. The role of the tympana. Hear Res 115:27–38

Bauer M, von Helversen O (1987) Separate localization of sound recognizing and sound producing neural mechanisms in a grasshopper. J Comp Physiol A 161:95–101

Benda J, Herz AVM (2003) A universal model for spikefrequency adaptation. Neural Comput 15:2523–2564

Boyan G, Williams L, Meier T (1993) Organization of the commissural fibers in the adult brain of the locust. J Comp Neurol 332:358–377

Bräunig P, Pflüger HJ (2001) The unpaired median neurons of insects. Adv Insect Physiol 28:185–266

Burrows M, Laurent G (1993) Synaptic potentials in the central terminals of locust proprioceptive afferents generated by other afferents from the same sense organ. J Neurosci 13:808–819

Dobler S, Stumpner A, Heller K-G (1994a) Sex-specific spectral tuning for the partner's song in the duetting bushcricket *Ancistrura nigrovittata* (Orthoptera: Phaneropteridae). J Comp Physiol A 175:303–310

Dobler S, Heller K-G, Helversen O (1994b) Song pattern recognition and an auditory time window in the female bushcricket *Ancistrura nigrovittata* (Orthoptera: Phaneropteridae). J Comp Physiol A 175:67–74

Faure PA, Hoy RR (2000a) Auditory symmetry analysis. J Exp Biol 203:3209–3223

Faure PA, Hoy RR (2000b) The sounds of silence: cessation of singing and song pausing are ultrasound-induced acoustic startle behaviors in the katydid *Neoconocephalus ensiger* (Orthoptera; Tettigoniidae). J Comp Physiol A 186:129–142

Field LH, Matheson T (1998) Chordotonal organs of insects. Adv Insect Physiol. 27:1–228

Hardt M (1988) Zur Phonotaxis von Laubheuschrecken: Eine vergleichend verhaltensphysiologisch/neuroanatomische Untersuchung. PhD-thesis, Universität Bochum

Hardt M, Watson AH (1999) Distribution of input and output synapses on the central branches of bushcricket and cricket auditory afferent neurones: immunocytochemical evidence for GABA and glutamate in different populations of presynaptic buttons. J Comp Neurol 403:281–294

Heinrich R, Jatho M, Kalmring K (1993) Acoustic transmission characteristics of the tympanal tracheae of bushcrickets (Tettigoniidae). II: comparative studies of the tracheae of seven species. J Acoust Soc Am 93:3481–3489

Heller K-G (1988) Bioakustik der europäischen Laubheuschrecken. Ökologie in Forschung und Anwendung Bd. 1:Verlag Josef Markgraf, Weihersheim

Heller K-G, von Helversen D (1986) Acoustic communication in phaneropterid bushcrickets: species-specific delay of female stridulatory response. Behav Ecol Sociobiol 18:189–198

Hennig M, Franz A, Stumpner A (2004) Processing of auditory information in insects. Microsc Res Tech 63:351–374

Hill KG, Oldfiel BP (1981) Auditory function in Tettigoniidae (Orthoptera, Ensifera). J Comp Physiol A 142:169–180

Höbel G, Schul J (2007) Listening for males and bats: spectral processing in the hearing organ of *Neoconocephalus bivocatus* (Orthoptera: Tettigoniidae). J Comp Physiol A 193:917–925

Hoffmann E, Jatho M (1995) The acoustic trachea of Tettigoniids as an exponential horn: theoretical calculations and bioacoustical measurements. J Acoust Soc Am 98:1845–1851

Horseman G, Huber F (1994) Sound localisation in crickets. I. Contralateral inhibition of an ascending auditory interneuron (AN1) in the cricket *Gryllus bimaculatus*. J Comp Physiol A 175:389–398

Hoyle G, Dagand D, Moberl B, Colquhoun W (1974) Dorsal unpaired median insect neurons make neurosecretory endings on skeletal muscle. J Exp Zool 187:159–165

Hummel J, Kössl M, Nowotny M (2011) Sound-induced tympanal membrane motion in bushcrickets and its relationship to sensory output. J Exp Biol 214:3596–3604
Imaizumi K, Pollack GS (2005) Central projections of auditory receptor neurons of crickets. J Comp Neurol 493:439–447
Kalmring K, Kühne R (1980) The coding of airborne-sound and vibration signals in bimodal ventral-cord neurons of the grasshopper *Tettigonia cantans*. J Comp Physiol 139:267–275
Kalmring K, Rössler W, Ebendt R, Ahi J, Lakes R (1993) The auditory receptor organs in the forelegs of bushcrickets: physiology, receptor cell arrangement, and morphology of the tympanal and intermediate organs of three closely related species. Zool Jb Physiol 97:75–94
Kalmring K, Rössler W, Unrast C (1994) Complex tibial organs in the forelegs, midlegs, and hindlegs of the bushcricket *Gampsocleis gratiosa* (Tettigoniidae): comparison of the physiology of the organs. J Exp Biol 270:155–161
Kalmring K, Sickmann T, Jatho M, Zhantiev R, Grossbach M (1997) The auditory-vibratory sensory system of *Polysarcus denticauda* (Phaneropterinae, Tettigoniidae): III. Physiology of the ventral cord neurons ascending to the head ganglia J Exp Zool 279:9–28
Korsunovskaya OS, Zhantiev RD (2007) Effect of temperature on auditory receptor functions in crickets (Orthoptera, Tettigoniodea). J Evol Biochem Physiol 43:327–334
Kostarakos K, Hennig M, Römer H (2009) Two matched filters and the evolution of mating signals in four species of cricket. Front Zool 6:22
Kühne R (1982) Neurophysiology of the vibration sense in locusts and bushcrickets: response characteristics of single receptor units. J Insect Physiol 28:155–163
Lakes-Harlan R, Bailey WW, Schikorski T (1991) The auditory system of an atympanate bushcricket *Phasmodes ranatriformes* (Westwood)(Tettigoniidae: Orthoptera). J Exp Biol 158:307–324
Lewis DB (1974) The physiology of the Tettigoniid ear. I. The implications of the anatomy of the ear to its function in sound reception. J Exp Biol 60:821–837
Libersat F, Hoy RR (1991) Ultrasonic startle behavior in bushcrickets (Orthoptera; Tettigoniidae). J Comp Physiol A 169(4):507–514
Lin Y, Kalmring K, Jatho M, Sickmann T, Rössler W (1993) Auditory receptor organs in the forelegs of *Gampsocleis gratiosa* (Tettigoniidae); Morphology and function of the organs in comparison to the frequency parameters of the conspecific song. J Exp Zool 267:377–388
Lynch JW, Rajendra S, Barry PH, Schofield PR (1995) Mutations affecting the glycine receptor agonist transduction mechanism convert the competitive agonist, picrotoxin, into an allosteric potentiator. J Biol Chem 270:13799–13806
Machens CK, Stemmler MB, Prinz P, Krahe R, Ronacher B, Herz AV (2001) Representation of acoustic communication signals by insect auditory receptor neurons. J Neurosci 21:3215–3227
Marshall DC, Hill KBR (2009) Versatile aggressive mimicry of cicadas by an Australian predatory katydid. PLoS ONE 4:e4185
Mason AC (1991) Hearing in a primitive ensiferan: the auditory system of Cyphoderris monstrosa (Orthoptera, Haglidae). J Comp Physiol A 168:351–363
Mason AC, Faure PA (2004) The physiology of insect auditory afferents. Microsc Res Tech 63:338–350
Mason AC, Morris GK, Hoy RR (1999) Peripheral frequency mismatch in the primitive ensiferan Cyphoderris monstrosa (Orthoptera:Haglidae). J Comp Physiol A 184:543–551
Michelsen A, Heller KG, Stumpner A, Rohrseitz K (1994) A new biophysical method to determine the gain of the acoustic trachea in bushcrickets. J Comp Physiol A 175:145–151
Michelsen A, Larsen ON (1978) Biophysics of the ensiferan ear. I. Tympanal vibrations in bushcrickets (Tettigoniidae) studied with laser vibrometry. J Comp Physiol 123:193–203
Molina J, Stumpner A (2005) Effects of pharmacological treatment and photoinactivation on the directional responses of an insect neuron. J Exp Zool 303A:1085–1103
Montealegre-Z F, Jonsson T, Robson-Brown KA, Postles M, Robert D (2012) Convergent evolution between insect and mammalian audition. Science 338(6109):968–971

Morris GK (1971) Aggression in male conocephaline grasshoppers (Tettigoniidae). Anim Behav 19:132–137

Nocke H (1975) Physical and physiological properties of the tettigoniid ("grasshopper") ear. J Comp Physiol 100:25–57

Nowotny M, Hummel J, Weber M, Möckel D, Kössl M (2010) Acoustic-induced motion of the bushcricket (*Mecopoda elongata*, Tettigoniidae) tympanum. J Comp Physiol A 196:939–945

Oldfield BP (1982) Tonotopic organisation of auditory receptors in Tettigoniidae (Orthoptera: Ensifera). J Comp Physiol 147:461–469

Oldfield BP (1983) Central projections of primary auditory fibres in Tettigoniidae (Orthoptera: Ensifera). J Comp Physiol 151:389–395

Oldfield BP (1984) Physiology of auditory receptors in two species of Tettigoniidae (Orthoptera: Ensifera). J Comp Physiol 155:689–696

Oldfield BP (1985) The role of the tympanal membranes in the tuning of auditory receptors in Tettigoniidae (Orthoptera: Ensifera). J Exp Biol 116:493–497

Oldfield BP, Hill KG (1983) The physiology of ascending auditory interneurons in the tettigoniid *Caedicia simplex* (Orthoptera: Ensifera): response properties and a model of integration in the afferent auditory pathway. J Comp Physiol 152:495–508

Ostrowski TD (2006) Adaptation von auditorischen Neuronen von Langfühlerschrecken (Ensifera). Diploma thesis, Universität Göttingen

Ostrowski TD (2009) Filtering of species specific song parameters via interneurons in a bush cricket's brain. PhD-thesis, Universität Göttingen

Ostrowski TD, Stumpner A (2010) Frequency processing at consecutive levels in the auditory system of bush crickets (Tettigoniidae). J. Comp. Neurol. 518:3101–3116

Palghat Udayashankar A, Kössl M, Nowotny M (2012) Tonotopically arranged traveling waves in the miniature hearing organ of bushcrickets. PLoS ONE 7:e31008

Pires A, Hoy RR (1992) Temperature coupling in cricket acoustic communication. I. Field and laboratory studies of temperature effects on calling song production and recognition in *Gryllus firmus*. J Comp Physiol A 171:69–78

Pollack GS (1988) Selective attention in an insect auditory neuron. J Neurosci 8:2635–2639

Pollack GS, Imaizumi K (1999) Neural analysis of sound frequency in insects. BioEssays 21:295–303

Popov AV, Markovich AM, Andjan AS (1978) Auditory interneurones in the prothoracic ganglion of the cricket, *Gryllus bimaculatus*. I. The large segmental auditory neuron (LSAN). J Comp Physiol 126:183–192

Poulet JFA, Hedwig B (2005) Auditory orientation in crickets: pattern recognition controls reactive steering. PNAS 102:15665–15669

Reeve R, Webb B (2003) New neural circuits for robot phonotaxis. Phil Trans R Soc Lond A (Math) 361:2245–2266

Rheinlaender J (1975) Transmission of acoustic information at three neuronal levels in the auditory system of *Decticus verrucivorus* (Tettigoniidae, Orthoptera). J Comp Physiol 97:1–53

Rheinlaender J, Kalmring K (1973) Die afferente Hörbahn im Bereich des Zentralnervensystems von *Decticus verrucivorus* (Tettigoniidae). J Comp Physiol 85:361–410

Rheinlaender J, Kalmring K, Römer H (1972) Akustische Neuronen mit T-Struktur im Bauchmark von Tettigoniiden. J Comp Physiol 77:208–224

Römer H (1983) Tonotopic organization of the auditory neuropile in the bushcricket, *Tettigonia viridissima*. Nature 306:60–62

Römer H (1985) Anatomical representation of frequency and intensity in the auditory system of orthoptera. In: Kalmring K, Elsner N (eds) Acoustic and vibrational communication in insects. Parey, Hamburg, pp 25–32

Römer H (1987) Representation of auditory distance within a central neuropil of the bushcricket *Mygalopsis marki*. J Comp Physiol A 161:33–42

Römer H (1993) Environmental and biological constraints for the evolution of long-range signalling and hearing in acoustic insects. Trans R Soc Lond B 226:179–185

Römer H, Bailey WJ (1986) Insect hearing in the field. II. Male spacing behaviour and correlated acoustic cues in the bushcricket *Mygalopsis marki*. J Comp Physiol A 159:627–638
Römer H, Krusch M (2000) A gain-control mechanism for processing of chorus sounds in the afferent auditory pathway of the bushcricket *Tettigonia viridissima* (Orthoptera; Tettigoniidae). J Comp Physiol A 186:181–191
Römer H, Rheinlaender J, Dronse R (1981) Intracellular studies on auditory processing in the metathoracic ganglion of the locust. J Comp Physiol 144:305–312
Römer H, Marquart V, Hardt M (1988) Organization of a sensory neuropile in the auditory pathway of two groups of Orthoptera. J Comp Neurol 275:201–215
Rössler W, Hübschen A, Schul J, Kalmring K (1994) Functional morphology of bushcricket ears: comparison between two species belonging to the Phaneropterinae and Decticinae (Insecta, Ensifera). Zoomorphol 114:39–46
Römer H, Hedwig B, Ott SR (2002) Contralateral inhibition as a sensory bias: the neural basis for a female preference in a synchronously calling bushcricket, *Mecopoda elongata*. Eur J Neurosci 15:1655–1662
Schildberger K (1984) Temporal selectivity of identified auditory neurons in the cricket's brain. J Comp Physiol A 155:171–185
Schul J (1997) Neuronal basis of phonotactic behaviour in *Tettigonia viridissima*: Processing of behavioural relevant signals by auditory afferents and thoracic interneurons. J Comp Physiol A 180:573–583
Schul J, Sheridan RA (2006) Auditory stream segregation in an insect. Neuroscience 138:1–4
Schul J, Matt F, von Helversen O (2000) Listening for bats: The hearing range of the bushcricket *Phaneroptera falcata* for bat echolocation calls measured in the field. Proc R Soc Lond B 267:1711–1715
Schul J, Mayo AM, Triblehorn JD (2012) Auditory change detection by a single neuron in an insect. J Comp Physiol A online. doi:10.1007/s00359-012-0740-3
Schulze W, Schul J (2001) Ultrasound avoidance behaviour in the bushcricket *Tettigonia viridissima* (Orthoptera: Tettigoniidae). J Exp Biol 204:733–740
Schumacher R (1975) Scanning-electron-microscope description of the tibial tympanal organ of the Tettigonioidea (Orthoptera, Ensifera). Z Morph Tiere 81:209–219
Schumacher R (1979) Zur funktionellen Morphologie des auditiven Systems der Laubheuschrecken (Orthoptera: Tettigoniidae). Entomol Gen 5:321–356
Schwabe J (1906) Beitrage zur Morphologie und Histologie der tympanalen Sinnesapparate der Orthopteren. Zoologica 20:1–154
Selverston AI, Kleindienst H-U, Huber F (1985) Synaptic connectivity between cricket auditory interneurons as studied by selective photoinactivation. J Neurosci 5:1283–1292
Seymour C, Lewis B, Larsen ON, Michelsen A (1978) Biophysics of the ensiferan ear. J Comp Physiol 123:205–216
Shen JX (1993a) A peripheral mechanism for auditory directionality in the bushcricket *Gampsocleis gratiosa*: acoustic tracheal system. J Acoust Soc Am 94:1211–1217
Shen JX (1993b) Morphology and physiology of auditory interneurons of the bushcricket *Gampsocleis gratiosa*. Jap J Physiol 43:S239–S246
Sickmann T (1997) Vergleichende funktionelle und anatomische Untersuchung zum Aufbau der Hör- und Vibrationsbahn im thorakalen Bauchmark von Laubheuschrecken. PhD Thesis, Philipps-Universität Marburg, Marburg
Siegert ME, Römer H, Hashim R, Hartbauer M (2011) Neuronal correlates of a preference for leading signals in the synchronizing bushcricket *Mecopoda elongata* (Orthoptera, Tettigoniidae). J Exp Biol 214:3924–3934
Sobel EC, Tank DW (1994) In vivo Ca^{2+}-dynamics in a cricket auditory neuron: an example of chemical computation. Science 263:823–826
Stölting H (1996) Morphologische und elektrophysiologische Charakterisierung auditorischer Rezeptoren bei *Pholidoptera griseoaptera* (Tettigoniidae, Decticinae). Diploma thesis, Universität Göttingen

Stölting H, Stumpner A (1998) Tonotopic organization of auditory receptorcells in the bushcricket *Pholidoptera griseoaptera* (Tettigoniidae, Decticini). Cell Tissue Res 294:377–386

Stradner J, Römer H (2008) Reliable coding of small behaviourally relevant interaural intensity differences in a pair of interneurons of an insect. Biol Lett 4:711–714

Strauß J, Lakes-Harlan R (2008a) Neuroanatomy of the complex tibial organ of *Stenopelmatus* (Orthoptera: Ensifera: Stenopelmatidae). J Comp Neurol 511:81–91

Strauß J, Lakes-Harlan R (2008b) Neuroanatomy and physiology of the complex tibial organ of an atympanate ensiferan, *Ametrus tibialis* (Brunner von Wattenwyl, 1888) (Gryllacrididae, Orthoptera) and evolutionary implications. Brain Behav Evol 71:167–180

Strauß J, Lakes-Harlan R (2009) The evolutionary origin of auditory receptors in Tettigonioidea: the complex tibial organ of Schizodactylidae. Naturwissenschaften 96:143–146

Stritih N, Stumpner A (2009) Vibratory interneurons in the non-hearing cave cricket indicate evolutionary origin of sound processing elements in Ensifera. Zoology 112:48–68

Stumpner A (1995) Some new local and descending auditory neurons in the prothoracic ganglion of crickets and bushcrickets. In: Elsner N, Menzel R (eds) Göttingen neurobiology report. Thieme, Stuttgart, p A274

Stumpner A (1996) Tonotopic organization of the hearing organ in a bushcricket—Physiological characterization and complete staining of auditory receptor cells. Naturwiss 83:81–84

Stumpner A (1997) An auditory interneurone tuned to the male song frequency in the duetting bushcricket *Ancistrura nigrovittata* (Orthoptera, Phaneropteridae). J Exp Biol 200:1089–1101

Stumpner A (1998) Picrotoxin eliminates frequency selectivity of an auditory interneuron in a bushcricket. J Neurophysiol 79:2408–2415

Stumpner A (1999) Comparison of morphology and physiology of two plurisegmental sound-activated interneurones in a bushcricket. J Comp Physiol A 185:199–205

Stumpner A (2002) A species-specific frequency filter through specific inhibition, not specific excitation. J Comp Physiol A 188:239–248

Stumpner A, Molina J (2006) Diversity of intersegmental auditory neurons in a bush cricket. J Comp Physiol A 192:1359–1376

Stumpner A, Atkins G, Stout J (1995) Processing of unilateral and bilateral auditory inputs by the ON1 and L1 interneurons of the cricket *Acheta domesticus* and comparison to other cricket species. J Comp Physiol A 177:379–388

Suga N, Katsuki Y (1961a) Central mechanism of hearing in insects. J Exp Biol 88:545–558

Suga N, Katsuki Y (1961b) Pharmacological studies on the auditory synapses of a grasshopper. J Exp Biol 88:759–770

ter Hofstede HM, Fullard JH (2008) The neuroethology of song cessation in response to gleaning bat calls in two species of katydids, *Neoconocephalus ensiger* and *Amblycorypha oblongifolia*. J Exp Biol 211:2431–2441

ter Hofstede HMT, Kalko EKV, Fullard JH (2010) Auditory-based defence against gleaning bats in neotropical katydids (Orthoptera: Tettigoniidae). J Comp Physiol A 196:349–358

Thompson KJ, Siegler MVS (1991) Anatomy and physiology of spiking local and intersegmental interneurons in the median neuroblast lineage of the grasshopper. J Comp Neurol 305:659–675

Triblehorn JD, Schul J (2009) Sensory-encoding differences contribute to species specific call recognition mechanisms. J Neurophysiol 102:1348–1357

Watson AHD (2002) Presynaptic modulation of sensory neurons in the segmental ganglia of arthropods. Microsc Res Tech 58:262–271

Zhantiev RD, Korsunovskaya OS (1978) Morpho-functional organization of tympanal organs in *Tettigonia cantans* (Orthoptera, Tettigoniidae). Zool Zh 57:1012–1016

Zhantiev RD, Korsunovskaya OS (1983) Structure and functions of two auditory neurons in the bush cricket *Tettigonia cantans* Fuess. (Orthoptera, Tettigoniidae). Revue d'Entomologie de L'URSS 62:462–469

Zhantiev RD, Korsunovskaya OS (1990) Sound communication of Phaneropteridae (Orthoptera). In: Gribakin FG et al (eds) Sensory systems and communication in arthropods. Basel, Birkhäuser, pp 402–406

Zhantiev RD, Korsunovskaya OS, Chukanov VS (2004) Effect of sound signals on spontaneous activity in grasshopper interneurons (Orthoptera, Tettigoniidae). J Evol Biochem Physiol 40:531–538

Zimmermann U, Rheinlaender J, Robinson D (1989) Cues for male phonotaxis in the duetting bushcricket *Leptophyes punctatissima*. J Comp Physiol A 164:621–628

Chapter 10
Evolution of Call Patterns and Pattern Recognition Mechanisms in *Neoconocephalus* Katydids

Johannes Schul, Sarah L. Bush and Katy H. Frederick

Abstract In the katydid genus *Neoconocephalus*, males typically produce continuous calls with an extremely fast pulse rate of about 200/s. Divergence from this ancestral pattern includes alternation of pulse periods resulting in a double-pulse pattern, and the grouping of pulses into chirps. Double-pulse patterns evolved five times independently in the genus. Analysis of the female preferences and call recognition mechanisms revealed that in three species with double-pulse pattern, females have independently evolved new mechanisms for recognizing the derived call pattern. In the remaining two species with double-pulse pattern, females retain the ancestral recognition mechanism and exhibit no preference for the derived temporal pattern. These results suggest that males are leading the evolutionary divergence of call patterns in this genus. We propose a hypothetical scenario in which genetic bottlenecks and founder effects arising from the climatic history of North America contributed to the rapid diversification of calls in this genus.

10.1 Introduction

The diversity of acoustic communication signals has long-fascinated naturalists and biologists of diverse interests. The evolutionary mechanisms generating and maintaining the diversity of signals and signal recognition mechanisms remain a controversial topic, despite a large body of research performed over the last four decades.

J. Schul (✉) · S. L. Bush · K. H. Frederick
Division of Biological Sciences, University of Missouri, 207 Tucker Hall, Columbia, MO 65211, USA
e-mail: schulj@missouri.edu

S. L. Bush
e-mail: bushsl@missouri.edu

K. H. Frederick
e-mail: khfrederick@gmail.com

B. Hedwig (ed.), *Insect Hearing and Acoustic Communication*,
Animal Signals and Communication 1, DOI: 10.1007/978-3-642-40462-7_10,

The importance of sexual selection and especially female choice for the diversification of communication signals has been widely recognized and well documented in many systems (Andersson and Simmons 2006). For example, in a radiation of Hawaiian crickets (genus *Laupala*), species differ significantly in the pulse rate of male calls, and female preferences for pulse rate provide an explanation for this diversity (Grace and Shaw 2011; Mendelson and Shaw 2005). Directional female preferences for certain call parameters that potentially encode male quality or condition (e.g., chirp/pulse rate, chirp duration, and carrier frequency) are common (Andersson 2006). However, among closely related species, such directional preferences typically point in the same direction (e.g., toward longer calls or lower carrier frequency) and thus are unlikely to contribute to call divergence. Accordingly, call traits under directional sexual selection typically are not divergent among closely related species.

In contrast, call traits that establish diversity, such as differences in call pattern or rhythm, are typically under stabilizing selection by female preferences (Helversen and Helversen 1994). Such a situation is less obviously explained by sexual selection models of signal evolution. To understand the evolutionary processes that lead to this divergence, comparative studies involving many species are necessary. We summarize here a series of studies on the evolution of a novel call character, the "double-pulse pattern" in *Neoconocephalus* (Orthoptera, Tettigoniidae). This call character evolved multiple times convergently in this group, and the phenotypes of female call recognition support a rarely considered model of divergence. We contrast the situation in *Neoconocephalus* with that in a similarly well studied group of Hawaiian crickets.

10.2 The Katydid Genus *Neoconocephalus*

The natural history and acoustic communication system has been reviewed in detail by Greenfield (1990). In short, the American genus *Neoconocephalus* is most likely the closest relative of the genus *Ruspolia*, which occurs throughout the western hemisphere. About 25 *Neoconocephalus* species exist in the Caribbean, Central and North America; the South American *Neoconocephalus* fauna has not been systematically revised and it is unclear how many additional species exist in this range (Greenfield 1990).

All *Neocononcephalus* species are grassland species with typical habitats ranging from marshes to prairies. Most species have wide areas of occurrence with overlapping ranges. Males produce calls by rubbing their forewings (elytra) against each other. Females are silent and, when receptive, walk or fly toward calling males (phonotaxis). It is common to hear up to four species calling at the same time in one habitat. Morphological diversity among the *Neoconocephalus* species is limited and the male calls are the main diagnostic character, in addition to the shape and coloration of the cone (fastigium) between the antennae (Walker and Greenfield 1983; Walker 2012).

All *Neoconocephalus* and *Ruspolia* species have long wings and are strong fliers. Long range dispersal has been observed in several species. For example, *Ruspolia* swarms have been observed on ships in the Atlantic, several hundreds of km from the African coast, supporting the hypothesis that the American *Neoconocephalus* fauna originated through colonization from African *Ruspolia* (reviewed in Greenfield 1990). We found calling *N. triops* more than 150 km north of their known range following the occurrence of strong low pressure weather systems. The high mobility predisposes *Neoconocephalus* to rapid colonization of new habitats (e.g., after glaciations) and thus might play an important role in the evolution of the call diversity.

10.3 Diversity of Male Calls in *Neoconocephalus*

Male calls in *Neoconocephalus* are unusual for Tettigoniids in two respects. First, the spectral energy is concentrated in a narrow low frequency band with center frequency typically at 10–15 kHz, and with ultrasonic components strongly attenuated (Schul and Patterson 2003). Second, the rate of the pulses produced by the repetitive opening and closing movements of the forewings is typically extremely high at 150–250 Hz (Greenfield 1990).

We conducted a Bayesian character state analysis of male call traits based on the molecular phylogeny of *Neoconocephalus* (Snyder et al. 2009). This analysis revealed that the ancestral call state is a continuous 'buzz' consisting of pulses repeated with a single, fast repetition rate (Fig. 10.1). Three derived call patterns occur in *Neoconocephalus*: (1) discontinuous calls, where pulses are grouped in a second order time structure (chirps, or verses), (2) a dramatic reduction of the pulse rate below 50 Hz, and (3) the presence of two alternating pulse periods, which results in pulse pairs or 'double-pulses.' Five species in *Neoconocephalus* have this double-pulse pattern. Quite astonishingly, this novel call trait occurs at five tips in the phylogeny (Fig. 10.1, arrows). Bayesian character state reconstruction confirmed that this novel call trait evolved at least five times independently in this group, providing an excellent opportunity to study the evolutionary processes underlying the diversification of the communication system.

The distribution of call diversity among *Neoconocephalus* species shows an interesting pattern of rapid evolutionary diversifications as well as phenotypic stability. Seven temperate species form one monophyletic clade (label T in Figs. 10.1, 10.2), which contains the same diversity in call traits as is found in the complete genus (Fig. 10.2). The sibling taxon to the temperate clade is the tropical species *N. triops*, which has an extremely large range from the southern USA to Argentina (Greenfield 1990). While pulse rate varies by about 20 % across this range, all populations studied so far have the derived double-pulse pattern. However, the genetic diversity among different Caribbean and Central American populations of *N. triops* is greater than that among the seven

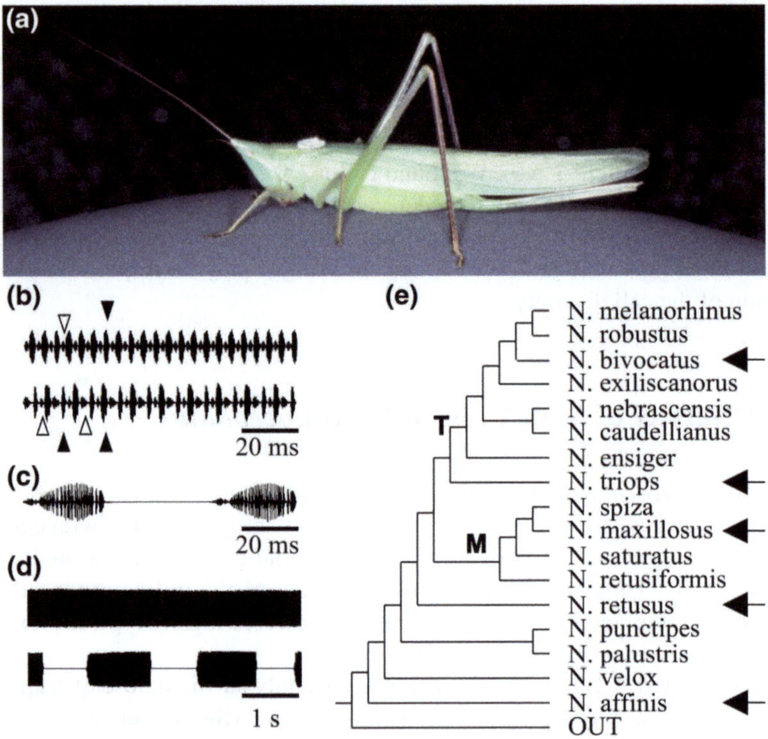

Fig. 10.1 Diversity of male calls in *Neoconocephalus* katydids. Three lines of call diversification occur in this genus (a–c). **a** Female *N. bivocatus* on the 'Kramer Kugel'. **b** Most species produce a single-pulse period repeated at extremely fast rates, with loud pulses produced during the closing movements of the tegmina (*top trace*), while five species have alternating pulse periods, resulting in a double-pulse pattern (*bottom trace*). **c** A second line of diversification is a dramatic reduction of pulse rate. Filled arrowheads indicate sounds produced during the closing movements of the tegmina, open triangles those during the opening movement. **d** Most species produce continuous calls (*top*), while some have added a second order time structure (verses or chirps) to their calls (*bottom*). **e** Phylogenetic relationship among 17 *Neoconocephalus* species; total evidence tree based on Amplified Fragment Length Polymorphisms (AFLP) and nuclear and mitochondrial DNA sequences (from Snyder et al. 2009). The support values of all in-group nodes are >0.98. Arrows denote the five species with calls with double-pulse pattern. T and M labels indicate the 'temperate' and '*N. maxillosus*' clades referenced in the text

species of the temperate clade, which have much higher phenotypic call diversity. Applying standard molecular clocks (Brower 1994) to the ultrametric tree (Fig. 10.2) suggests that the radiation of the temperate species occurred within the last 100,000 years, i.e., within the last glacial cycle. Similarly short divergence times appear in another branch of the phylogeny (the *N. maxillosus* clade, M in Figs. 10.1, 10.2), which is similar in call diversity. The genus thus includes examples of both stability of call phenotypes despite genetic diversity in *N. triops*, and call diversification despite little genetic diversity within the temperate and *N. maxillosus* clades.

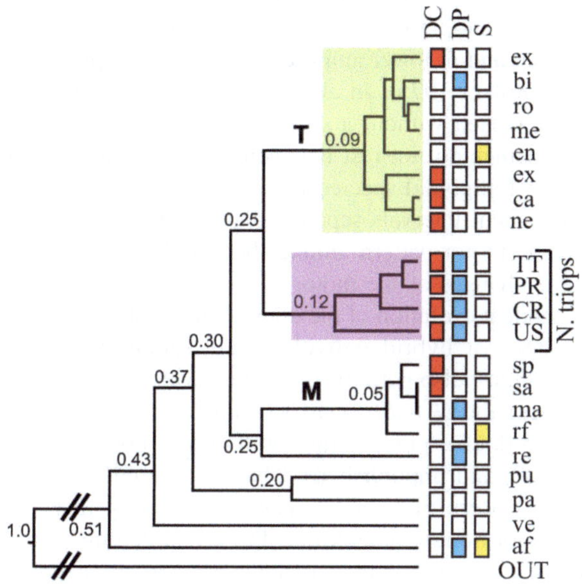

Fig. 10.2 Relative time tree of the genus *Neoconocephalus*. The branch lengths indicate the relative timing of divergence with the numbers indicating the relative age of the nodes. Estimates of the divergence time of the temperate clade (*green shading*) range from 18,000 to 72,000 years, suggesting that the call divergence in this clade evolved during the last glacial cycle (for methods see below). The species are labeled as in Fig. 10.1. For *N. triops* (purple shading), four populations are given separately (*TT* Trinidad, *PR* Puerto Rico, *CR* Costa Rica, *US* continental USA). The columns on the right show the presence of the three derived call traits in each species (*DC*, discontinuous calls, *red*; *DP* double pulses, *blue*; *S* slow pulse rate, *yellow*) with filled squares indicating the derived state and open squares the ancestral state (continuous calls, single-pulse pattern, fast pulse rate, respectively). The labels indicate the first two letters of the species name. **T** and **M** indicate the 'temperate' and '*N. maxillosus*' clades referenced in the text (*Methods:* Tree is based on mitochondrial DNA sequences (CO1, 875 bp), which are included in the total evidence tree in Fig. 10.1. It was built using Bayesian methods in BEAST (v. 1.72; Drummond et al. 2006; Drummond and Rambaut 2007). Divergence times were estimated using BEAST and r8 s (v. 1.71; Sanderson 2003). Mutation rates were assumed in the range from 1.1 to 3.83 %/per million years.)

10.4 Double-Pulse Pattern: A Novel Signal Trait in *Neoconocephalus*

In species with the ancestral 'single-pulse pattern', males call by opening and closing their forewings without interruptions: the wings do not rest in the closed position before the new cycle begins. The derived double-pulse pattern in *Neoconocephalus* is produced by four movements: a full opening of the forewings, a partial closing, a partial re-opening, and a full closing (Fig. 10.3; Walker 1975). The two sound pulses produced during the closing movements have alternating pulse periods and typically different durations and amplitudes: the pulses produced during the partial closing movement are shorter and of lower amplitude than the ones produced during the full closure (Fig. 10.3; Walker 1975). In all five species with double-pulse pattern, the double-pulses are repeated without a silent interval occurring between them, indicating that the forewings do not rest between the double-pulses. This is markedly different from other Tettigoniid species with double-pulses, where wing movement stops and extended silent intervals separate double-pulses (Fig. 10.3; Heller 1988). The similarity of the motor pattern among the five *Neoconocephalus* species suggests that the same changes of the motor pattern (and probably similar mutations) underlie the independent evolution of the double-pulse pattern.

Anecdotal findings of hybrid individuals give additional clues regarding the evolution of the double-pulse pattern. Hybrid individuals between *N. bivocatus* (a double-pulse species) and *N. nebrascensis* (a single-pulse species with discontinuous calls) produce a double-pulse pattern within the verse structure of *N. nebrascensis* (Büttner 2002, and unpublished). Thus, the double-pulse pattern does not mix with the verse pattern of *N. nebrascensis*, indicating that double-pulses and verse structure are not homologous traits. Thus, double-pulses are a modification of the single-pulse pattern, rather than an added second order time structure.

In addition to the five occurrences in *Neoconocephalus*, double-pulsed calls also evolved in many other Tettigoniid genera, e.g., *Tettigonia*, *Platycleis*, *Metrioptera*, *Decorana*, *Sepiana*, *Tessellana* (Heller 1988), *Atlanticus* (Walker 1975), and *Orchelimum* (Thomas and Alexander 1962). Double-pulse rhythms are also found in other signal modalities, e.g., in the vibratory signals of the katydid genus *Meconem* (Heller 1988), or in the blinking patterns of courting fireflies (Coleoptera: Lampyridae; Lloyd 1984). The propensity of *Neoconocephalus* species to evolve the double-pulse pattern, and the prevalence of such double-pulse rhythms across communication modalities and phylogenetic groups, suggests a predisposition of basic neuronal rhythm generators to switch from a single-pulse rhythm to this derived pattern. If such a predisposition exists, it is plausible that only a small number of mutations—potentially even a single mutation—might be required for this novel pattern to arise.

10.5 Female Call Recognition and Preferences

To the human ear, *Neoconocephalus* calls with fast pulse rates and single-pulse patterns sound continuous without amplitude modulation. Due to nonlinearities in our auditory signal transduction, we perceive a fast pulse rate of about 200 Hz as the pitch of a pure tone underlying the high pitched carrier signal (McDermott and Oxenham 2008).

In species with a fast single-pulse pattern, female preferences reflect human perception: Females respond to continuous signals without amplitude modulation (Fig. 10.4a), but are highly selective against signals that contain silent intervals longer than a few ms (Deily and Schul 2004, 2009). Females are not selective for

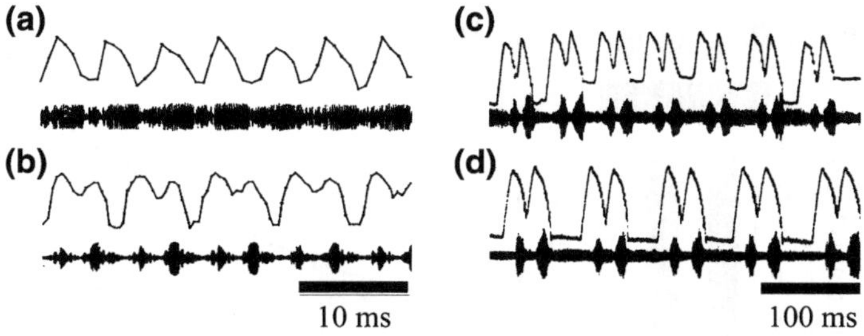

Fig. 10.3 Forewing movements (*top trace*) and sound pulses (*bottom trace*) during call production of **a** *N. robustus*, a species with single-pulse pattern, and three species with double-pulse pattern: **b** *N. bivocatus*, **c** *Tettigonia viridissima*, **d** *Metrioptera oblongicollis*. In *N. bivocatus* the wings do not stop between double-pulses, while in other Tettigoniid species the wings rest between double-pulses. Note the differences in time scale between **a**, **b** and **c**, **d**. From Walker (1975) (**a**, **b**) and Heller (1988) (**c**, **d**)

any particular temporal pattern (e.g., pulse rate or pulse duration), as long as the signal does not contain longer silent intervals. While this call recognition mechanism does not select for a specific temporal pattern, it forces males to maintain a fast pulse rate that is effectively perceived as a continuous signal. It is worth noting that the female call recognition mechanism accepts a much wider range of the signal parameter space (defined by pulse and interval duration) than is occupied by the conspecific male calls (Fig. 10.4b). This does not mean that females are unselective, as they are highly selective for the interval duration. However, the large accepted parameter space provides the potential for diversification of the temporal call pattern, without losing attractiveness (Deily and Schul 2004).

This mechanism in which females respond to calls that lack silent gaps has been described in five species (Greenfield 1993; Deily and Schul 2004, 2009; Bush and Schul 2010) and occurs in at least two more (*N. exciliscanorus*, *N. palustris*, unpublished data). The distribution of this recognition mechanism in the phylogeny of *Neoconocephalus* makes it very likely that it represents the ancestral state. In species with a verse or chirp pattern (*N. nebrascensis*, *N. exciliscanorus*) an additional mechanism is involved in recognizing the second order time structure (Deily and Schul 2009, unpublished results).

10.5.1 Recognition of Calls with Double-Pulse Pattern: Three Species with Derived Mechanisms

Male calls of *N. bivocatus* and *N. triops* have fast double-pulse rates of 80–110/s, i.e., they maintained the ancestral fast pulse rate of approximately 200 wingstrokes/s. Females of both species recognize the conspecific rate of the double-pulses: they respond with phonotaxis only if the rate of the double-pulses is in the correct range (Deily and Schul 2004; Beckers and Schul 2008). The

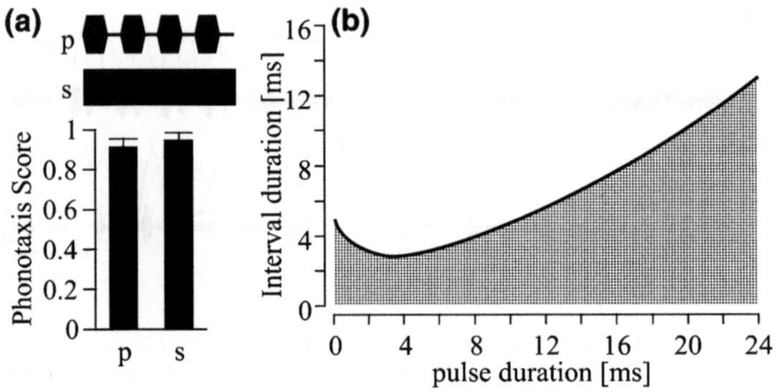

Fig. 10.4 Call recognition in female *N. robustus*, a species with fast single-pulse calls. **a** Phonotaxis scores (mean ± s.e.m., $n = 10$) in response to a call model consisting of pulses (p) and an unmodulated sine wave (s). **b** Response field of *N. robustus* females to combinations of pulse interval and duration. The gray area denotes the range of stimulus parameters that elicit a significant phonotactic response. After Deily and Schul (2004)

actual double-pulse structure of the call is not necessary, as double-pulses can be replaced by a single pulse of appropriate length. Females respond to the correct (double-) pulse rate in a wide range of duty cycles (Fig. 10.5a,b). Although preferences were similar in these two species, they appear to differ in the mechanisms generating them. In *N. triops*, call recognition is most likely generated by a neuronal oscillation: females show significant responses to half the double-pulse rate and to stimuli with rhythmically altered intervals (Fig. 10.5c). In *N. bivocatus*, in contrast, we could not detect either of these responses (Fig. 10.5c), suggesting that a different neuronal mechanism (e.g., temporal integration) underlies rate recognition (see Bush and Schul 2006 for a detailed discussion of the experimental logic).

In both species, females respond to a distinctly different pattern than is generated by the conspecific males: females recognize a single rate (i.e., the repetition rate of the double-pulses) while males produce pulses with two alternating rates. This "nonparallel" evolution of male calls and female recognition (Schul and Bush 2002) is a strong argument against genetic coupling of the sender and receiver subunits of the communication system (Butlin and Ritchie 1989; Bloake 1991), which was recently implicated in the diversification of a cricket communication system (Shaw and Lesnick 2009; Wiley and Shaw 2010). In *Neoconocephalus*, it is unlikely that the same mutation(s) would alter the call pattern generator toward double-pulses in males and at the same time halve the preferred pulse rate in females. Furthermore, as described below, the change in male call is not always accompanied by a change in female call recognition. The nonparallel changes of sender and receiver indicate that they are mainly linked through the function of the communication system, rather than by shared genetics. This raises the question of how the match between sender and receiver is assured during the evolution of the novel calls and recognition mechanisms. We propose an evolutionary scenario in the final section of this chapter.

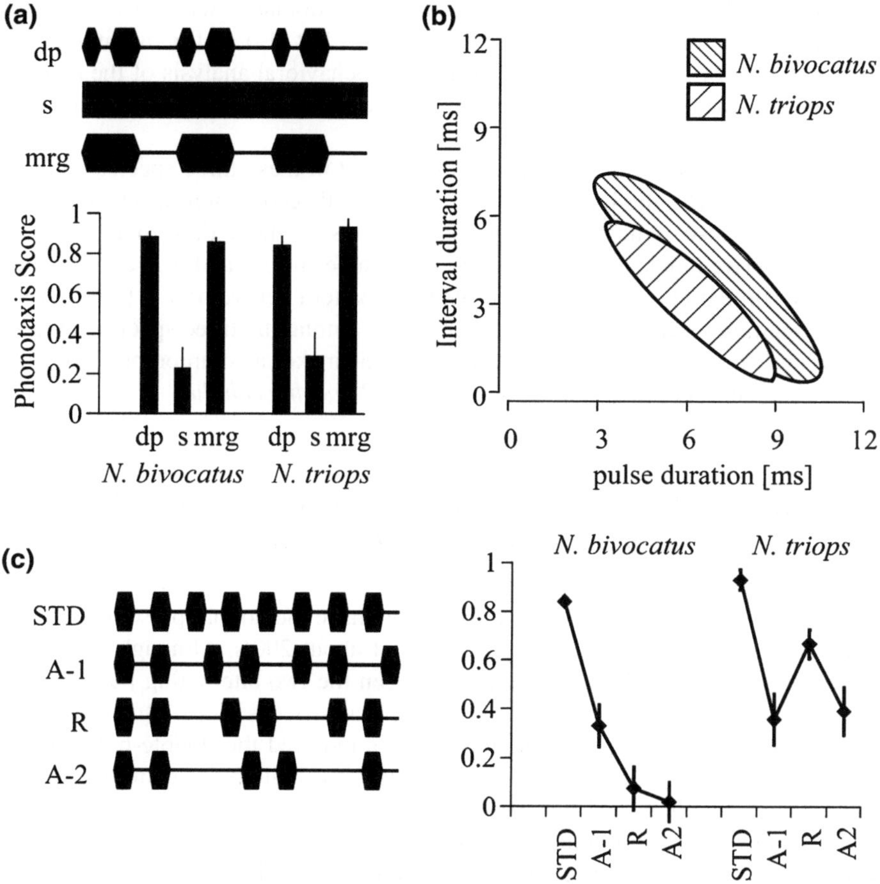

Fig. 10.5 Call recognition in female *N. bivocatus* and *N. triops*, two species with double-pulse pattern. **a** Phonotaxis scores in response to call models with the conspecific double-pulse pattern (dp), an unmodulated sine wave (s), and a model with the double-pulses merged into one long pulse (mrg). **b** Response fields: the gray area denotes the range of stimulus parameters that elicit significant responses. **c** Phonotaxis scores (mean ± s.e.m, $n = 8$–10) to stimuli in which every second interval was stretched, either rhythmically (r) so that all pulses fell at times where pulses occur in the standard model (std), or arrhythmically (a-1, a-2), so that half of the pulses occurred during silent intervals in the standard model. Strong responses to the rhythmical stimulus (r) indicate rate recognition based on neuronal resonance. Data for a, b taken from Deily and Schul (2004), and Beckers and Schul (2008)

Male calls of *N. affinis* also have a double-pulse pattern, however, the wingstroke rate is much slower with approximately 24 wingstrokes/s or 12 double-pulses/s. Female *N. affinis* evaluate the double-pulse rate and respond when it is near the conspecific value. In contrast to the previous two species, the double-pulse pattern is necessary and double-pulses cannot be replaced by one long pulse Fig. 10.6. Extensive testing (Bush et al. 2009) revealed that females recognize the amplitude modulation (AM) rate in the frequency domain (Schmidt et al. 2008) and require not only the rate of the double-pulses, but also its first harmonic (i.e., the wingstroke or single-pulse rate). A detailed behavioral analysis of the recognition mechanisms indicated that neuronal resonance underlies the recognition of at least the fundamental AM rate (Bush et al. 2009).

The double-pulse pattern of male calls evolved in these three species independently. Accordingly, the mechanisms of female call recognition differ among the three species, also indicating independent origins of these derived preferences. While the derived male pattern is similar in all three species and thus appears to be caused by similar changes in the neuronal call pattern generator, the three derived call recognition mechanisms differ qualitatively among the three species, suggesting major differences in auditory processing. This makes a common predisposition for the recognition of double-pulses unlikely in *Neoconocephalus*.

10.5.2 Recognition of Calls with Double-Pulse Pattern: Two Species Retain the Ancestral Mechanisms

Male calls of *N. retusus* and *N. maxillosus* both have the double-pulse pattern with wingstroke rates close to the ancestral state of about 200/s. Although the pattern is clearly double-pulsed, the difference between the two alternating pulse periods is less pronounced than in the three species previously described (Bush and Schul 2010). Surprisingly, females of neither species required the double-pulse pattern for call recognition but responded well to calls with single-pulse pattern or to continuous signals without AM Fig. 10.7a. Females responded well when the intervals between pulses were short enough; signals with longer intervals were unattractive, independent of the pulse rate (Bush and Schul 2010). Thus, female call recognition was identical to that of species with single-pulse pattern, i.e., it remains in the ancestral state, while male calls have a derived temporal pattern. Furthermore, females did not show preferences for either their derived conspecific pattern or the ancestral single-pulse pattern, as the intervals between double-pulses are short enough for female call recognition Fig. 10.7b (Bush and Schul 2010).

This pattern in which females are in the ancestral state and males are in the derived state means that in these two species, males are leading the evolutionary divergence of the communication system and that female choice is not the driving force. Other factors linked to female choice, such as localizability or the transmission of the preferred pattern through the habitat, do not play a role here: localizability was included in the evaluation of the female preferences, and degradation of the pulse pattern does not interfere with the ancestral call recognition mechanism, as echoes would mask silent intervals (Bush and Schul 2010). Male–male interactions in *Neoconocephalus* take place at the chirp, but not at the pulse level (Meixner and Shaw 1986; Greenfield 1983, 2005), at time scales an order of magnitude longer than the pulse pattern in these species. Thus, intrasexual competition also fails to explain the novel call pattern here.

Natural selection may also have profound influence on signal traits. Eavesdropping predators/parasitoids and energetic costs may lead to the evolution of novel traits (Zuk and Kolluru 1998). *Neoconocephalus* katydids are commonly hosts of acoustically orienting tachinid flies. The double-pulse pattern seemingly does not provide an advantage, as species with double-pulse pattern are among the heavily parasitized species; infection rates of 75–90 % occur both in single and double-pulse species (Burk 1982; Talwar 2007, see Chap. 4 by Hedwig and Robert). The double-pulse pattern also has the same number of pulses as an equivalent single-pulse pattern and should have similar energetic cost. If there were selection to save energy during calling, male *N. retusus* or *N. maxillosus* could have reduced the pulse rate while maintaining the single-pulse pattern; double-pulses introduce longer silent intervals and thus impose a faster minimum pulse rate (=wing stroke rate) than single-pulses to remain attractive. As discussed in detail by Bush and Schul (2010), neither sexual nor natural selection provide convincing explanations for the evolution of the double-pulse pattern in *N. retusus* and *N. maxillosus*. This leaves the hypothesis of neutral evolution, i.e., that the novel call pattern evolved through mutations and genetic drift. It is important to bear in mind that neutral evolution is the evolutionary null hypothesis, which can only be supported by the absence of evidence for alternative explanations. Of course, one can never exclude the possibility that selection acted on this trait in the past but is no longer detectable. In the absence of convincing arguments for selective advantages of the double-pulse pattern, however, neutral evolution is the most parsimonious explanation.

10.6 A Scenario for the Evolution of Double-Pulse Pattern in *Neoconocephalus*

We develop here a scenario for the evolution of the novel call traits and recognition mechanisms in *Neoconocephalus* based on the findings that (1) the double-pulse pattern evolved five times independently, (2) the females of three species have derived call recognition mechanisms, whereas (3) in two species, female call recognition remains in the ancestral state and no convincing adaptive explanation for the double-pulse pattern exists.

We propose that the few mutations required to change the male call pattern to a double-pulse pattern happened as the first step, with female recognition remaining in the ancestral state. As long as this mutant/derived call pattern still falls within the attractive range of female call recognition, i.e., if the intervals between double-pulses are short enough to be tolerated, it will not reduce the attractiveness of the derived calls (Fig. 10.4). Thus, the new call pattern would be selectively neutral and could persist in the population, resulting in populations in which some males produce the ancestral single-pulse pattern and others the new double-pulse pattern.

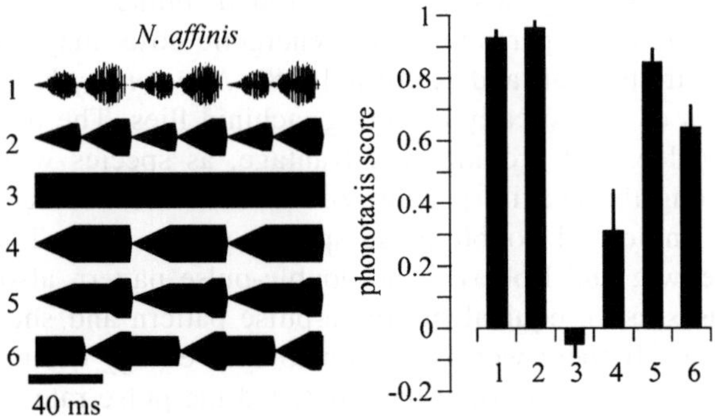

Fig. 10.6 Call recognition in female *N. affinis*, a species with double-pulse pattern and a much slower pulse rate than typical for *Neoconocephalus* (approximately 12 double pulses/s). Phonotaxis scores (mean ± s.e.m., $n = 8$–11) to a recording (*1*) and model (*2*) of a conspecific call were high, while a continuous sine wave (*3*) did not elicit responses. A model with the double-pulse replaced by one long pulse was also unattractive (*4*); adding a second amplitude modulation during this long pulse (*5, 6*) restored the attractiveness of the model. Thus, the double-pulse structure is necessary for female responses, which differs from the other two species shown in Fig. 10.5. Data from Bush et al. (2009)

The likelihood of a neutral trait becoming fixed in a large population is low. The ecological history of North America, however, suggests that founder events, genetic bottlenecks, and population fragmentation have likely been frequent occurrences in the history of this genus, greatly increasing the probability of fixation of a neutral trait through genetic drift. Throughout the Pleistocene, ranges of *Neoconocephalus* have changed dramatically as a result of repeated glacial—interglacial cycles. During glaciations, habitats suitable for *Neoconocephalus* in North America were restricted to small stretches near the Gulf of Mexico (Adams 1997). The Neotropics were much drier, such that ranges of tropical *Neoconocephalus* were certainly very different from the ranges that exist today. During interglacials, suitable habitats spread through the Caribbean and eastern parts of North America. Strong flight muscles and a capacity for long range dispersal provided opportunities for rapid range expansion of *Neoconocephalus* as habitats became suitable.

Even on a much shorter time scale, *Neoconocephalus* distributions are fluid. Over the past 12 years, we have observed two dramatic changes in population sizes, including one local near-extinction of a *N. robustus* population and a large-scale regional drop by >90 % of *N. nebrascensis*. In both cases, population levels returned to previous sizes within a few years. Founder and bottleneck effects are therefore likely to have been common during the diversification of *Neoconocephalus*.

Given their high mobility, it is conceivable that new populations were frequently founded by few individuals. If the founders of such new populations had the mutation(s) for a novel call trait, a population with the derived call pattern would be established and have the opportunity to diverge at least somewhat in other traits, before secondary contact with populations in the ancestral state

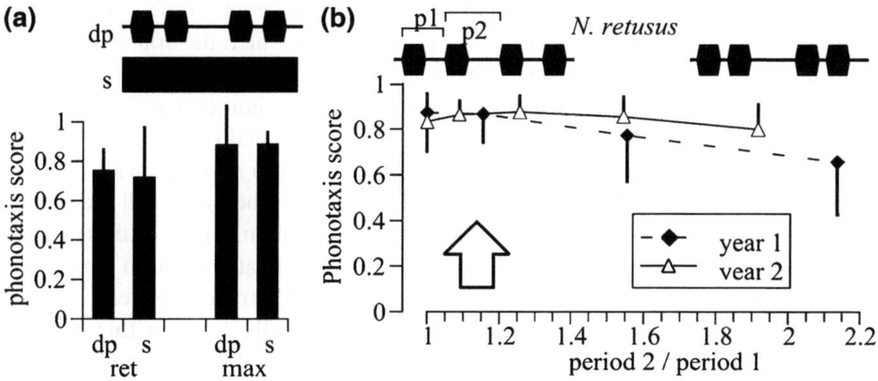

Fig. 10.7 Call recognition and preferences in female *N. retusus* and *N. maxillosus*, two species with double-pulse pattern. **a** Phonotaxis scores (mean ± 95 % confidence interval, $n = 6$–8) in response to call models with double-pulses (dp) and unmodulated sine waves (s). Strong responses to the sine wave resemble the situation in single-pulse species (Fig. 10.4) rather than that in double-pulse species (Figs. 10.5, 10.6). **b** Absence of female preference for the double-pulse structure. Phonotaxis scores (mean ± 95 % CI, $n = 9$–10) of *N. retusus* females toward calls that vary from single-pulses to extreme double-pulses as measured by the ratio of period 2/period 1 (p2/p1, see inset). The experiment was run in two consecutive years with different females. The arrow indicates the mean ratio found in natural male calls

occurred. We hypothesize that *N. maxillosus* and *N. retusus* originated in this manner and that the communication system has remained in this state with derived male calls and ancestral female call recognition.

Mutations that influence female call recognition inevitably occur, too. While we generally think of female preferences as quantitative traits which evolve gradually, recent advances in computational neuroscience have demonstrated the qualitative effect that small changes in ion channel composition may cause. For example, adding or replacing a single channel may change the response type of a neuron from integrator to resonator (or vice versa) (Izhikevich 2001). Thus, even a single mutation might have a significant impact on call recognition.

In most cases, mutations of the call recognition mechanism that lead to significant changes of selectivity would not respond to the conspecific (ancestral) call pattern and would thus quickly disappear from the population. In a population in which the derived call trait (e.g., double-pulses) is either common or fixed, there are two scenarios by which a mutation in the female recognition system may spread. First, if the new recognition mechanism does respond well enough to a calling male to elicit a female approach, then founder and bottleneck effects could cause fixation (perhaps followed by refinement) of the derived recognition mechanism. Given the dispersal capabilities of *Neoconocephalus* and the quickly changing habitats at the end of glaciation, even a single female with the derived recognition mechanism could establish a new, isolated population as these insects spread into newly available habitats following the end of an ice age.

An alternative scenario involves selection rather than genetic drift. The mutation of the recognition system may appear in a population in which the double-pulse pattern is already fixed but call recognition remains in the ancestral state (i.e., similar to *N. retusus* or *N. maxillosus*). When this population comes into contact with a different species with single-pulse calls, females with the novel call recognition (responding to double-pulses only) would have an advantage over females with ancestral recognition, as they would avoid heterospecific matings. The novel call recognition would thus be selected for and spread through the population. Either of these scenarios would result in a population fixed both in a novel call pattern and in a novel call recognition mechanism that prefers the novel over the ancestral call pattern.

Given the apparent ease with which double-pulse calls arose in this genus, it is possible that double-pulses evolved and disappeared multiple times in both the single- and double-pulse species throughout history. The new call trait would have been stabilized by a novel female preference, while in the absence of a change in female preference, the call trait may be lost again through drift. In this respect, the current situation in *N. maxillosus* and *N. retusus*, with males in the derived state and females in the ancestral state, may be temporary in evolutionary time.

While we propose that the double-pulse pattern evolved in *Neoconocephalus* through genetic drift, sexual selection has clearly influenced other call parameters within the genus. For example, the call of *N. robustus* has a significantly lower carrier frequency than its congeners, probably as a result of selection favoring females that avoided heterospecific matings with *N. bivocatus* (Deily and Schul 2004, 2006). In addition, the female preference for leading males in *N. spiza* (Greenfield and Roizen 1993; Greenfield and Schul 2008) may have arisen as a result of benefits obtained by females who mate with leading males.

10.7 Comparison to Hawaiian Crickets

The co-evolution of male calls and female preferences has been studied in detail in the Hawaiian cricket genus *Laupala* (Orthoptera, Gryllidae). Here, male calls vary widely in pulse rate among species, and female preferences are tuned to conspecific pulse rates, functioning as premating isolation barriers (Grace and Shaw 2011; Shaw and Herlihy 2000). Many small effect mutations likely underlie this diversity (Ellison et al. 2011). Indirect (Shaw and Lesnick 2009; Wiley and Shaw 2010) and direct (Wiley et al. 2012) evidence indicates that male calls and female preferences are genetically linked through multiple loci, assuring the function of the communication system during diversification. This is in stark contrast to *Neoconocephalus*, where call production and call recognition seemingly evolved much more independently and may involve few mutations with large effects.

Differences in ecology between *Neoconocephalus* and *Laupala* provide a potential explanation for the contrasting evolutionary scenarios. *Laupala* divergence occurred across the Hawaiian islands over the last 4–6 million years in a relatively stable tropical environment, with newly appearing islands providing space for diversification. In contrast, repeated cycles of ice-ages and interglacials dramatically altered the climate and habitats where *Neoconocephalus* evolved and occurs today. These different climatic conditions might favor different evolutionary processes resulting in contrasting phenotypic and genetic patterns. We acknowledge that we have not yet identified the genetic basis underlying the diversity of the communication system in *Neoconocephalus*, but rather infer our hypothetical scenario from a large collection of phenotypic, phylogenetic and ecological data. Additional studies including the genetic as well as neuronal basis of *Neoconocephalus* communication are ultimately needed to provide a complete and conclusive picture. At the current stage, *Neoconocephalus* serves as a reminder that evolutionary patterns are complex and might lead to surprising explanations.

10.8 Conclusion

Among *Neoconocephalus* katydids, diversity in the communication system consists largely of qualitative changes in both male calls and female recognition mechanisms. In some species, changes in male calls have preceded changes in female recognition, suggesting that males rather than females may lead the divergence of the communication system, and that sender and receiver may evolve more independently of each other than is commonly assumed. The rate of diversification since the last glacial cycle suggests that genetic drift (e.g., founder effects) may have contributed to the radiation within this genus. Further research is needed to identify the mechanisms of divergence, including potential differences in patterns of divergence between organisms living in tropical versus temperate ecological zones.

Acknowledgments We thank the numerous members of our lab for their contributions and support. Oli Beckers, Josh Deily, and Rob Snyder each contributed significantly to this research. Tom Walker and Michael Greenfield introduced JS and SLB to many aspects of *Neoconocephalus* natural history and helped locate many species for these studies. We would like to thank Peter Heinecke for his help in assembling and maintaining our walking compensator ("Kramer Kugel"), which has been instrumental for this work.

References

Adams J (1997) Preliminary vegetation maps of the world since the last glacial maximum: an aid to archaeological understanding. J of Archaeol Sci 24:623–647

Andersson M (2006) Condition-dependent indicators in sexual selection: development of theory and tests. In: Lucas JR, Simmons LW (eds) Essays in animal behaviour: celebrating 50 years of animal behaviour. Elsevier, Amsterdam, pp 253–267

Andersson M, Simmons LW (2006) Sexual selection and mate choice. Trends Ecol Evol (Amst) 21:296–302. doi:10.1016/j.tree.2006.03.015

Beckers OM, Schul J (2008) Developmental plasticity of mating calls enables acoustic communication in diverse environments. Proc R Soc Lond B 275:1243–1248

Bloake CRB (1991) Coevolution of sender and receiver of sexual signals: genetic coupling and genetic correlations. Trends Ecol Evol 6:225–227. doi:10.1016/0169-5347(91)90027-U

Brower AVZ (1994) Rapid morphological radiation and convergence among races of the butterfly *Heliconius erato* inferred from patterns of mitochondrial DNA evolution. Proc Natl Acad Sci USA 91:6491–6495

Burk TE (1982) Evolutionary significance of predation on sexually signaling males. Fla Entomol 65:90–104

Bush SL, Schul J (2006) Pulse-rate recognition in an insect: evidence of a role for oscillatory neurons. J Comp Physiol A 192:113–121. doi:10.1007/s00359-005-0053-x

Bush SL, Schul J (2010) Evolution of novel signal traits in the absence of female preferences in *Neoconocephalus* katydids (Orthoptera, Tettigoniidae). PLoS ONE 5(8):e12457

Bush SL, Beckers OM, Schul J (2009) A complex mechanism of call recognition in the katydid *Neoconocephalus affinis* (Orthoptera: Tettigoniidae). J Exp Biol 212:648–655. doi:10.1242/jeb.024786

Butlin R, Ritchie MG (1989) Genetic coupling in mate recognition systems: what is the evidence. Biol J Linn Soc 37:237–246. doi:10.1111/j.1095-8312.1989.tb01902.x

Büttner UK (2002) Charakterisierung der Gesänge von fünf in Missouri (USA) heimischen *Neoconocephalus*-Arten (Orthoptera, Tettigoniidae). Diplom Thesis, Friedrich Alexander University, Erlangen

Deily JA, Schul J (2004) Recognition of calls with exceptionally fast pulse rates: Female phonotaxis in the genus *Neoconocephalus* (Orthoptera: Tettigoniidae). J Exp Biol 207:3523–3529

Deily JA, Schul J (2006) Spectral selectivity during phonotaxis: a comparative study in *Neoconocephalus* (Orthoptera, Tettigoniidae). J Exp Biol 209:1757–1764

Deily JA, Schul J (2009) Selective phonotaxis in *Neoconocephalus nebrascensis* (Orthoptera: Tettigoniidae): call recognition at two temporal scales. J Comp Physiol A 195:31–37

Drummond AJ, Rambaut A (2007) BEAST: bayesian evolutionary analysis by sampling trees. BMC Evol Biol 7:214

Drummond AJ, Ho SYW, Phillips MJ, Rambaut A (2006) Relaxed phylogenetics and dating with confidence. PLoS Biol 4:e88

Ellison CK, Wiley C, Shaw KL (2011) The genetics of speciation: genes of small effect underlie sexual isolation in the Hawaiian cricket *Laupala*. J Evolution Biol 24:1110–1119. doi:10.1111/j.1420-9101.2011.02244.x

Grace JL, Shaw KL (2011) Coevolution of male mating signal and female preference during early lineage divergence ofthe Hawaiian cricket, Laupala cerasina. Evolution 65:2184–2196. doi:10.1111/j.1558-5646.2011.01278.x

Greenfield MD (1983) Unsynchronized chorusing in the coneheaded katydid *Neoconocephalus affinis* (Beauvois). Anim Behav 31:102–112

Greenfield MD (1990) Evolution of acoustic communication in the genus *Neoconocephalus*: discontinuous songs, synchrony, and interspecific interaction. In: Bailey W, Rentz D (eds) The Tettigoniidae: biology, systematics, and evolution. Springer, New York, pp 72–97

Greenfield MD (1993) Inhibition of male calling by heterospecific signals: artifact of chorusing or abstinence during suppression of female phonotaxis? Naturwissenschaften 80:570–573

Greenfield MD (2005) Mechanisms and evolution of communal sexual displays in arthropods and anurans. Adv Study Behav 35:1–62

Greenfield MD, Roizen I (1993) Katydid synchronous chorusing is an evolutionarily stable outcome of female choice. Nature 364:618–620

Greenfield MD, Schul J (2008) Mechanisms and evolution of synchronous chorusing: emergent properties and adaptive functions in *Neoconocephalus* katydids (Orthoptera: Tettigoniidae). J Comp Psychology 122:289–297

Heller KG (1988) Bioakustik der Europäischen Laubheuschrecken. Margraf, Weikersheim

Ov Helversen, Dv Helversen (1994) Forces driving coevolution of song and song recognition in grasshoppers. In: Schildberger K, Elsner N (eds) Neural basis of behavioural adaptations. Fischer, Stuttgart, pp 253–284

Izhikevich EM (2001) Resonate-and-fire neurons. Neural Netw 14:883–894
Lloyd JE (1984) Evolution of a firefly flash code. Florida Entomol 67:228–239
McDermott JH, Oxenham AJ (2008) Music perception, pitch, and the auditory system. Curr Opin Neurobiol 18:452–463. doi:10.1016/j.conb.2008.09.005
Mendelson T, Shaw K (2005) Rapid speciation in an arthropod. Nature 433:375–376. doi:10.1038/433375a
Meixner AJ, Shaw KC (1986) Acoustic and associated behavior of the coneheaded katydid *Neoconocephalus nebrascensis* (Orthoptera: Tettigoniidae). Annals Entomol Soc Am 79:554–565
Sanderson MJ (2003) r8 s: inferring absolute rates of molecular evolution and divergence times in the absence of a molecular clock. Bioinformatics 19:301–302. doi:10.1093/bioinformatics/19.2.301
Schmidt A, Ronacher B, Hennig RM (2008) The role of frequency, phase and time for processing of amplitude modulated signals by grasshoppers. J Comp Physiol A 194:221–233. doi:10.1007/s00359-007-0295-x
Schul J, Bush SL (2002) Non-parallel coevolution of sender and receiver in the acoustic communication system of treefrogs. Proc R Soc Lond B 269:1847–1852
Schul J, Patterson AC (2003) What determines the tuning of hearing organs and the frequency of calls? a comparative study in the katydid genus *Neoconocephalus* (Orthoptera, Tettigoniidae). J Exp Biol 206:141–152
Shaw KL, Herlihy DP (2000) Acoustic preference functions and song variability in the Hawaiian cricket *Laupala cerasina*. Proc Biol Sci 267:577–584. doi:10.1098/rspb.2000.1040
Shaw KL, Lesnick SC (2009) Genomic linkage of male song and female acoustic preference QTL underlying a rapid species radiation. P Natl Acad Sci Usa 106:9737–9742. doi:10.1073/pnas.0900229106
Snyder RL, Frederick-Hudson KH, Schul J (2009) Molecular phylogenetics of the genus *Neoconocephalus* (Orthoptera, Tettigoniidae) and the evolution of temperate life histories. PLoS ONE 4:e7203. doi:10.1371/journal.pone.0007203.t001
Talwar M (2007) Function and evolution of the call spectrum in the katydid genus *Neoconocephalus*. Dissertation, University of Missouri, Columbia
Thomas ES, Alexander RD (1962) Systematics and behavioral studies on the meadow grasshoppers of the *Orchelimum concinnum* group (Orthoptera: Tettigoniidae). Occas Pap Mus Zool Univ Mich 626:1–31
Walker TJ (1975) Stridulatory movements of eight species of *Neoconocephalus* (Tettigoniidae). J Insect Physiol 21:595–603
Walker TJ (2012) The singing insects of North America—Katydids. http://buzz.ifas.ufl.edu/crickets.htm. Cited 1 June 2012
Walker T, Greenfield M (1983) Songs and systematics of Caribbean *Neoconocephalus* (Tettigoniidae, Orthoptera). Trans Amer Ent Soc 109:357–389
Wiley C, Ellison CK, Shaw KL (2012) Widespread genetic linkage of mating signals and preferences in the Hawaiian cricket Laupala. P Roy Soc B-Biol Sci 279:1203–1209. doi:10.1098/rspb.2011.1740
Wiley C, Shaw KL (2010) Multiple genetic linkages between female preference and male signal in rapidly speciating Hawaiian crickets. Evolution 64:2238–2245. doi:10.1111/j.1558-5646.2010.01007.x
Zuk M, Kolluru GR (1998) Exploitation of sexual signals by predators and parasitoids. Q Rev Biol 73:415–438

Chapter 11
Processing of Species-Specific Signals in the Auditory Pathway of Grasshoppers

Bernhard Ronacher

Abstract Species-specific acoustic signals of grasshoppers serve to attract mates; they are pivotal in avoiding hybridisation with sympatric species and to evaluate a potential mate's quality. This necessitates a high precision of neuronal processing, which is constrained by the noisy nature of neuronal activity. Applying a spike train metric to estimate the variability of auditory responses, we quantified the respective impacts that external degradation of acoustic signals and intrinsic neuronal noise exert on signal processing. Unexpectedly, the variability of spike patterns increases from the afferents to the neurons whose axons ascend to the brain and reduces their ability to discriminate between similar communication signals. Between thoracic local and ascending neurons a change of coding principles seems to occur, leading to a population code with labelled-line characteristics. Thoracic auditory processing is conserved between distantly related species, suggesting that during evolution the communication signals have been adapted to match properties of the receiver's sensory system.

11.1 Introduction

Many species of acridid grasshoppers produce acoustic signals in the context of mate attraction, and this communication is pivotal to avoid hybridisation with sympatric species (von Helversen and von Helversen 1975a, b; Stumpner and von Helversen 1994; Gottsberger and Mayer 2007). Hence, a high premium lies upon a reliable neuronal processing of these signals, allowing for their identification and classification. What cues of acoustic signals do grasshoppers use to infer species identity? The production as well as the recognition of songs is a genetically inherited, species-specific trait (von Helversen and von Helversen 1975a, b). Grasshopper ears exhibit only a poorly developed frequency resolution (Römer

B. Ronacher (✉)
Department of Biology and Bernstein Center for Computational Neuroscience Berlin,
Humboldt-Universität zu Berlin, Unter den Linden 6, 10099 Berlin, Germany
e-mail: bernhard.ronacher@rz.hu-berlin.de

B. Hedwig (ed.), *Insect Hearing and Acoustic Communication*,
Animal Signals and Communication 1, DOI: 10.1007/978-3-642-40462-7_11,

1976; Stumpner and von Helversen 2001; Hennig et al. 2004). In addition, the frequency spectra of songs, generated by rubbing the hind legs against the front wings, overlap considerably between species (Meyer and Elsner 1996). Hence it is unlikely that frequency analysis contributes substantially to species identification. Indeed, the major differences between the songs of different species reside in the species-specific patterns of amplitude modulations produced by the characteristic patterns of hind leg movements underlying sound production (e.g. Elsner 1974; Elsner and Popov 1978; von Helversen 1986; Hedwig 1992, 1994; Stumpner and von Helversen 1994; Mayer et al. 2010). In a pioneering study on *Chorthippus biguttulus* Dagmar von Helversen showed that it is indeed the pattern of amplitude modulations, i.e. the sound envelope, which provides the major cues for species identification (von Helversen 1972; see also Stumpner and von Helversen 1992; von Helversen and von Helversen 1994, 1997, 1998; Klappert and Reinhold 2003). *C. biguttulus* females respond in a band-pass like manner to simple song models that are composed of block-like sound syllables separated by pauses (Fig. 11.1b), and they seem to evaluate the ratio between syllable and pause durations (von Helversen 1979; von Helversen and von Helversen 1994).

There is evidence that specific features of acoustic signals may also be used in the context of *sexual selection*, i.e. for evaluating a potential mate's quality (Kriegbaum 1989; Kriegbaum and von Helversen 1992; Klappert and Reinhold 2003, 2005; Einhäupl et al. 2011; Stange and Ronacher 2012). To achieve this, the grasshoppers' auditory system must be able to detect subtle differences between songs of conspecifics which is obviously more difficult than to discriminate against the dissimilar signals of other species. We may therefore expect that the auditory system of these grasshoppers is particularly well adapted to process species-specific sound patterns. Indeed, behavioural experiments revealed an astonishing temporal resolution of the grasshoppers' auditory system: *C. biguttulus* females reject the songs of males that have lost one hind leg. The syllables of these males contain short silent gaps—which in intact males do not occur due to a phase shift between the movements of left and right hind leg (Elsner 1974). The specific rejection of such gappy songs demonstrates that grasshopper females detect gaps as small as 2–3 ms (Fig. 11.1; von Helversen 1972, 1979; von Helversen and von Helversen 1997). Thus, their auditory system resolves very fast amplitude modulations, and with respect to gap detection these insects are not inferior to most vertebrates (for a detailed comparison see Prinz and Ronacher 2002). Hence, grasshoppers became a well-established model system for investigating the processing of acoustic signals in a small nervous system.

The songs' temporal pattern of amplitude modulations is crucial for signal recognition and mate choice in grasshoppers. Starting from the observation of a remarkable gap detection capacity, I will focus in this review on the following questions:

(1) How are the acoustic signals represented in the auditory system and how do the animals achieve such a high temporal resolution with a small number of auditory neurons and in spite of rather unreliable neuronal signals?

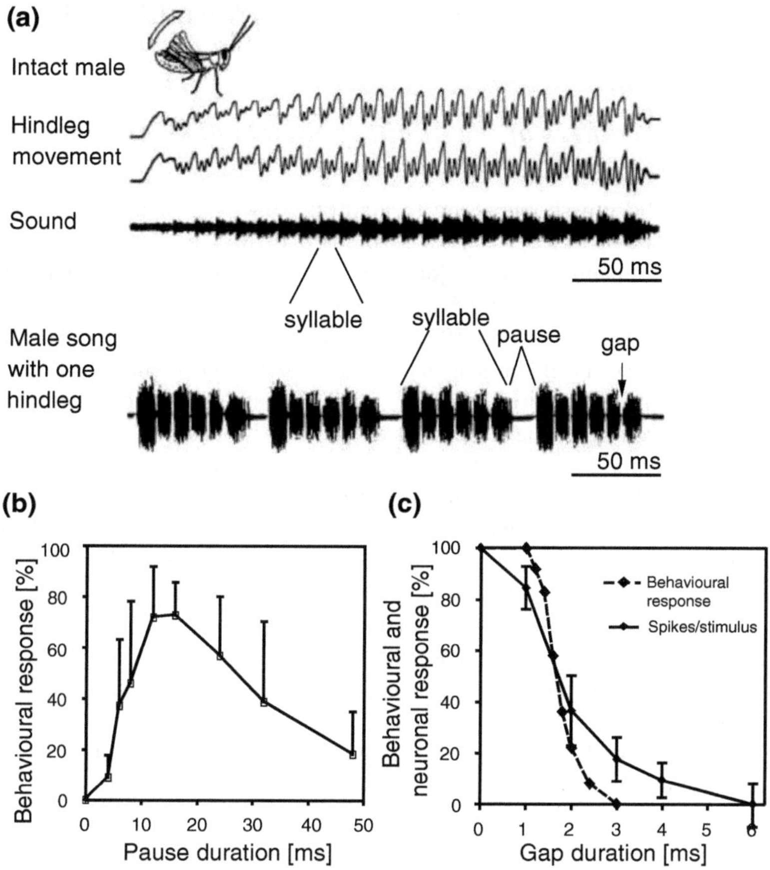

Fig. 11.1 **a** Hindleg movements and song pattern of an intact *Chorthippus biguttulus* male (*top*) and detail of a male song produced with only one hind leg (*bottom*). Since sound is not produced at the *upper* and *lower* reversal points of the leg movement, small gaps result if only one leg is used for sound production (*lower trace* in **a**). In the syllables of intact males there are no silent gaps due to a phase shift between the movements of the two legs. **b** Responses of ten females to model songs (*rectangularly* modulated broadband noise, syllable duration 80 ms). **c** Response of a female to song models containing gaps of different durations (*stippled curve*) and relative responses of three AN4-neurons to these stimuli. **a** combined from Ronacher et al. (2004), Krahe and Ronacher (1993); **b** Sträter and Ronacher unpublished; **c** combined after von Helversen (1972) and Franz and Ronacher (2002)

(2) How well can similar signals be discriminated—a question that is particularly relevant in the context of sexual selection? What coding strategies are used at different stages of the auditory pathway?
(3) How are the acoustic signals and the receiver's auditory pathway mutually adapted to each other, to guarantee an effective communication?

11.2 High Temporal Resolution in Spite of Variable Spike Trains

11.2.1 Basic Features of the Auditory System of Grasshoppers

The ears of acridid grasshoppers are located on both sides of the first abdominal segment. About 60–80 receptor cells assembled in Müller's organ attach to the inside of each tympanum. The majority of them is tuned to the range between 4 and 7 kHz, only around 10–15 afferent neurons are tuned to high frequencies, above 15 kHz (Michelsen 1971; Römer 1976; Halex et al. 1988; Jacobs et al. 1999). The axons of the sensory afferents enter the metathoracic ganglion complex and form an auditory neuropil with thoracic interneurons (Fig. 11.2a). Several interneuron types have been identified on the basis of their characteristic morphologies and activity patterns (Römer and Marquart 1984; Stumpner and Ronacher 1991). This auditory neuropil harbours an important stage of processing (Ronacher et al. 1986; Stumpner and Ronacher 1991; Stumpner et al. 1991). The basic organisation seems to be a 3-layer feed-forward network with afferent neurons connected to about 15 types of local neurons which in turn forward activity to about 20 different ascending neurons that send their axons to the brain (Marquart 1985; Römer et al. 1988; Stumpner 1988; Stumpner and Ronacher 1991; Boyan 1999; Vogel and Ronacher 2007). At least some afferent axons ascend even to the meso- and prothoracic ganglia (Halex et al. 1988; Römer et al. 1988; Stumpner 1988; Stumpner and Ronacher 1991), however, lesion experiments showed that

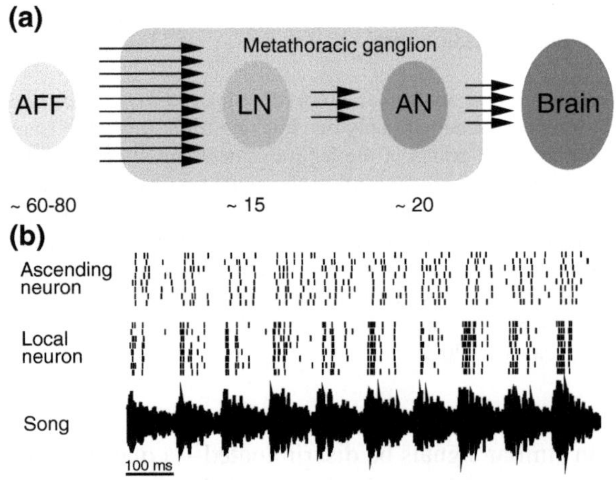

Fig. 11.2 **a** Schematic diagram of the auditory pathway of grasshoppers. *Aff* primary afferent neurons, *LN* local neurons, *AN* neurons with an axon ascending to the brain. **b** Spike raster plot of the responses of a local neuron and an ascending neuron to eight repeated presentations of an identical stimulus, a section of a *C. biguttulus* male song. Each *small bar* represents the timing of an action potential. **b** from Ronacher et al. (2008), with permission

these anterior projections of the sensory neurons are not necessary for song recognition (Ronacher et al. 1986). The spike patterns of ascending interneurons thus are the only activity which is forwarded to auditory networks in the brain and available to decide whether or not to accept a mating signal (Bauer and von Helversen 1987). A variety of physiological response types exists among thoracic interneurons, ranging from tonic, phasic-tonic and strongly phasic responses with different combinations of inhibition and excitation patterns (Römer and Marquart 1984, Römer et al. 1988; Stumpner and Ronacher 1991, Stumpner et al. 1991; Vogel et al. 2005; Vogel and Ronacher 2007).

For any recognition of acoustic signals based on the processing of amplitude modulations, the precision and reliability of a neuron's spike patterns is fundamental. It is a general observation that auditory neurons do not produce identical responses if one and the same sensory stimulus is presented repeatedly. The responses vary in the timing of spikes, in spike count, or in both (Fig. 11.2b). This trial-to-trial variability results from stochastic events at different stages of the sensory pathway, e.g. stochastic ion channel openings during sensory transduction and spike generation and stochastic transmitter release at synapses (e.g. Zador 1998; White et al. 2000). For the central nervous system that has to decide how the animal should react to objects and events in the outer world this "intrinsic" variability of spike trains poses a potentially difficult problem: if two spike trains differ to some extent, do they still represent the same object or different ones? To pursue this question we must take the view point of the central nervous system that has no other information about the surrounding world than the spike trains provided by sensory neurons. In other words, the question of whether two similar acoustic signals can be discriminated or not, converts to the question: are the sensory spike trains corresponding to the two acoustic signals sufficiently different to be distinguishable for down-stream neurons?

11.2.2 Comparing the Respective Impacts of Intrinsic and External Noise Sources

We first have to consider how relevant the intrinsic neuronal noise really is for the animals. As a rule, acoustic signals arriving at the receiver's ears will often be severely degraded compared to the emitted signal, due to a variety of external noise sources acting on sound transmission in the biotope (e.g. Michelsen and Larsen 1983; Lang 2000; Römer 2001; Brumm and Slabbekoorn 2005; Kostarakos and Römer 2010; Schmidt and Römer 2011). Conceivably, the intrinsic noise may play only a secondary role when we compare it with a strong degradation of acoustic signals occurring on the way between sender and receiver?

To compare the respective impacts of intrinsic neuronal noise versus external signal degradation we stimulated grasshopper males with a female song whose envelope was degraded by different amounts of random amplitude fluctuations (Fig. 11.3a). We quantified the dissimilarities between spike trains by means of a spike train metric (van Rossum 2001) and compared the spike trains that were elicited by the normal song with the spike trains produced in response

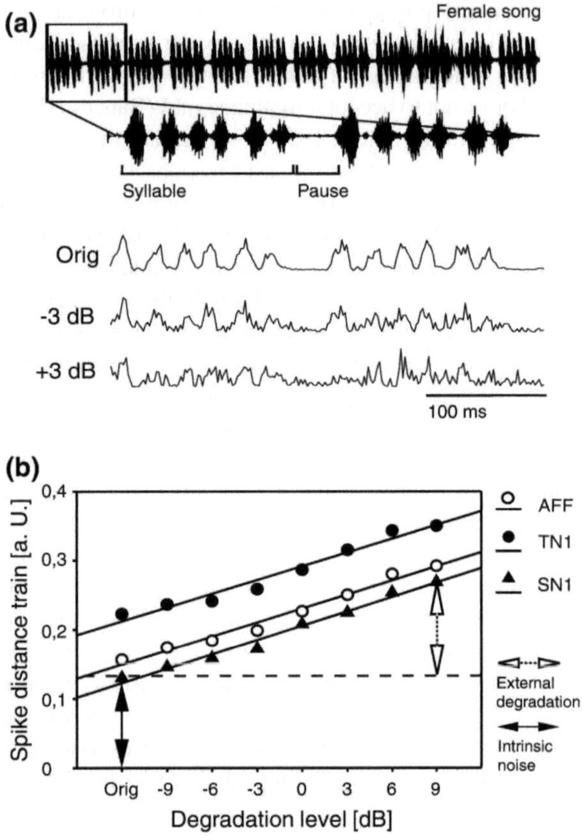

Fig. 11.3 **a** Song of a *C. biguttulus* female (*top trace*) and detail of two subunits. *Lower traces*: envelope of two original syllables and envelopes at two degradation levels (from Neuhofer et al. 2011). **b** Metric distances between spike trains obtained with the van Rossum metric at a resolution of $\tau = 5$ ms. For details of the procedure see Neuhofer et al. (2011). *Filled arrow* indicates the distance due to intrinsic noise, *open arrow* the additional distance caused by the external stimulus degradation. Data of an auditory afferent and two thoracic neurons (Neuhofer et al. unpublished)

to increasingly degraded songs (Neuhofer et al. 2011). The spike train metric describes the dissimilarity between two spike trains by a single number—small values indicating a high similarity (van Rossum 2001; Ronacher et al. 2008; Ronacher and Stange 2012). As the results for an auditory afferent and two thoracic interneurons show (Fig. 11.3b) the spike train distances increased linearly with increasing degradation level but they started already at a rather high distance value even for the normal stimulus (marked with orig at the abscissa). This distance value (filled arrow) reflects the trial-to-trial variation of the spike trains in response to repeated presentations of an identical stimulus. Most remarkably, even at the highest degradation levels the contribution of external signal degradation to the total spike train distance (open arrow in Fig. 11.3b) was not larger than the intrinsic distance. These examples clearly demonstrate that the intrinsic neuronal noise cannot be neglected in relation to the signal degradation imposed by external noise from the environment. Using this experimental approach, our original intention was to establish a kind of 'noise titration' by which we could determine that amount of external acoustic noise that would be equivalent to the contribution of intrinsic neuronal noise. However, in many neurons this 'titration' was not possible since the intrinsic noise was too dominant (Neuhofer et al. 2011). This was the case in most ascending neurons, for which the contribution of intrinsic noise to spike train distance was particularly strong compared to the additional distance introduced by external signal degradation (see Fig. 11.2 in Neuhofer et al. 2011).

We now can ask how this strong influence of intrinsic noise may affect the representation of acoustic signals in the auditory pathway, and in particular the resolution of fast amplitude modulations which are crucial for signal recognition as well as discriminating signals in the context of mate choice. The precision of spiking will influence the detection of subtle changes between signals—and thus the discrimination of similar songs—as well as the limits of temporal resolution, which will be discussed next.

11.2.3 Modulation Transfer Functions Only Partly Reveal the Temporal Resolution Capacities of Auditory Neurons

A widely applied method to measure the temporal resolution of time varying stimuli by sensory systems or by behaving animals is the modulation transfer function (MTF) paradigm (for reviews see Viemeister and Plack 1993; Joris et al. 2004). Using stimuli with sinusoidal amplitude modulations of different frequencies one can figure out if there are specific modulation frequencies to which the system responds particularly well, and what range of modulation frequencies the system is able to represent (Fig. 11.4a).

Applying the MTF paradigm to neuronal data one can focus on two variables: rate-MTF (rMTF) evaluate spike rates measured over a longer time period, while temporal-MTF (tMTF) describe how well spikes are phase locked to the stimulus envelope (Fig. 11.4b, c). rMTF of auditory neurons cover a wide range between all-pass, band-pass and band-stop characteristics (Weschke and Ronacher 2008; Wohlgemuth et al. 2011). Focusing on the question how well auditory neurons may resolve fast amplitude modulations, the tMTF paradigm reveals how well spikes are locked to the stimulus envelope in period histograms. The vector strength of the response is calculated and the upper limit of temporal resolution is described by the corner frequency, i.e. the upper limit up to which the spike pattern still follows the amplitude modulations (details in Prinz and Ronacher 2002; Weschke and Ronacher 2008; Wohlgemuth et al. 2011). Examples for the vector strength-based tMTF of an afferent, a local, and an ascending interneuron are shown in Fig. 11.4c. The two auditory interneurons show both a band-pass characteristic in their tMTF, but differ strongly in their corner frequencies, indicated by the vertical line. When the corner frequencies of a large sample of neurons belonging to different processing stages are compiled (Fig. 11.4d) an interesting picture is revealed: afferents and local neurons with primary-like responses have a high temporal resolution capacity, median corner frequencies are around 150 Hz. Other local neurons exhibit a shift to a somewhat reduced temporal resolution (median 131 Hz) but operate still up to the same frequency range. However, the ascending neurons differ markedly by occupying a much lower range of modulation frequencies (median 48 Hz). This graph thus reveals a major result: the upper limits of

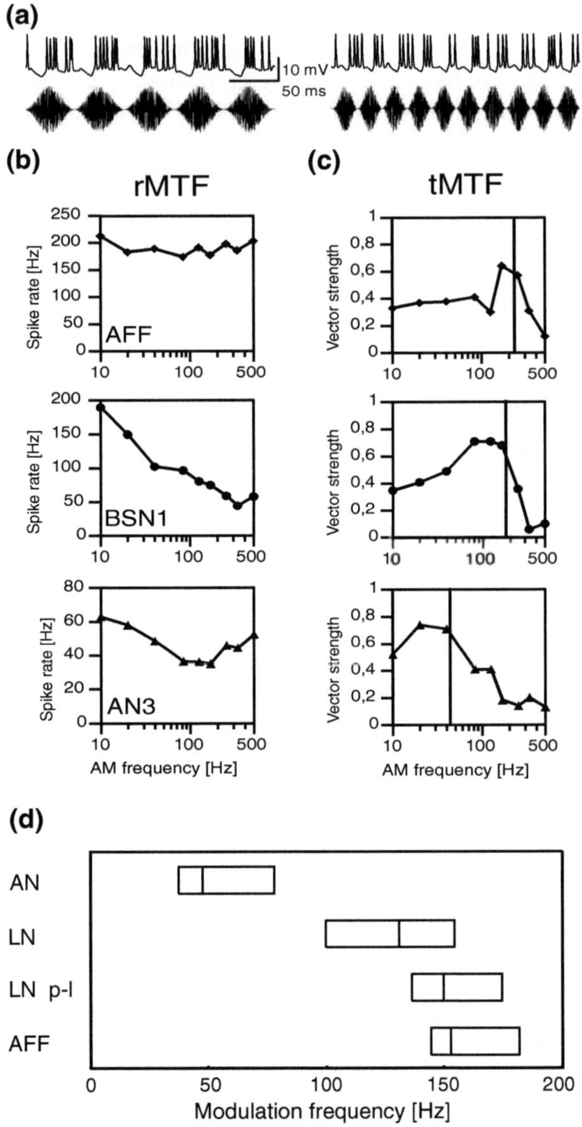

Fig. 11.4 Assessing temporal resolution with modulation transfer functions (MTF). **a** Response of a local neuron (BSN1) to a broadband noise stimulus with sinusoidal amplitude modulations of 20 and 40 Hz. **b, c** Rate and temporal MTFs of an afferent neuron, a local and an ascending neuron; *vertical lines* in **c** indicate the respective corner frequencies (for details see Wohlgemuth and Ronacher 2007). **d** Corner frequencies of auditory afferents ($N = 14$), local neurons of the primary-like response type ($N = 24$), other local neurons ($N = 18$) and ascending neurons ($N = 28$). Boxes indicate medians and quartile ranges; the corner frequencies of ANs are different from all others ($p < 0.001$), whereas all other data do not differ significantly (Kruskal–Wallis). **a–c** from Ronacher et al. (2008), with permission; **d** data combined from Weschke and Ronacher (2008) and Wohlgemuth et al. (2011)

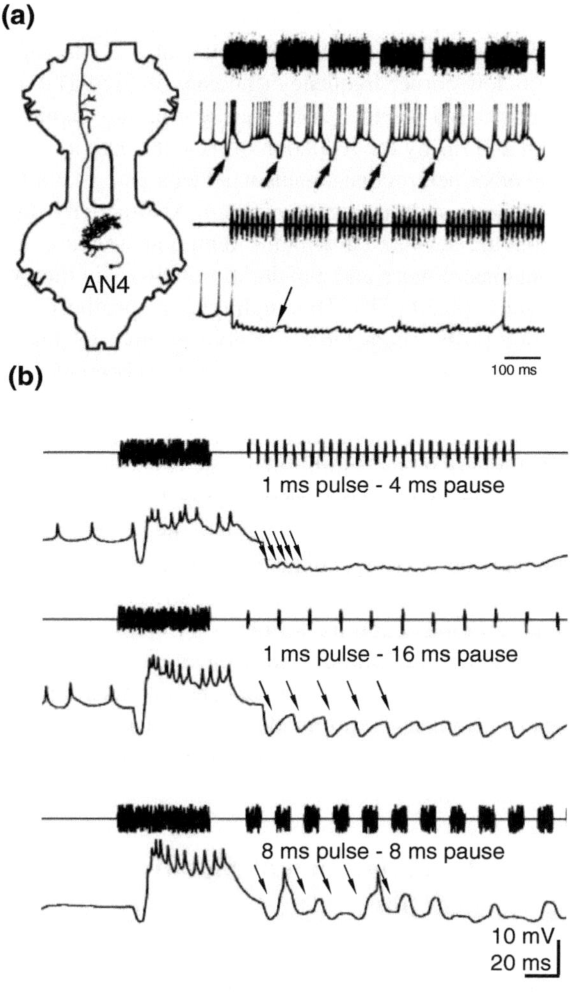

Fig. 11.5 Responses of the AN4 neuron to noise stimuli with silent gaps. **a** Morphology of the neuron and responses to a song model with uninterrupted syllables (*upper traces*) or syllables with 5 ms gaps (*lower*); *arrows* point at the inhibitory postsynaptic potentials (IPSPs) elicited by sound onsets. **b** Responses of an AN4 to different combinations of sound pulses and pause durations. **a** from Ronacher and Stumpner (1988), **b** A. Vogel unpublished results

temporal resolution are drastically reduced at the level of ascending neurons, as their median corner frequency of ~50 Hz corresponds to a time scale of 20 ms. This reduction of temporal resolution at the level of ascending neurons—which are the bottleneck for the information transfer to the brain—seems at odds with the capacity of behaving animals to detect gaps of 2–3 ms duration (see Fig. 11.1b).

How can we reconcile this specific behavioural response of the grasshopper females with the low corner frequencies of ascending neurons? The solution is found in the characteristic response type of an ascending neuron AN4 (Ronacher and Stumpner 1988). This neuron responds to a stimulus onset first with a pronounced IPSP preceding the spike response (arrows in Fig. 11.5a). However, if a continuous sound stimulus is interrupted by small gaps, this IPSP is triggered anew by each steep intensity rise, which leads to a quite efficient suppression of the neuron's spike response (bottom trace in Fig. 11.5a). When tested with different gap durations the spike activity of the AN4 neuron closely parallels the behaviour of females, with a substantial reduction of spike count at 2–3 ms gap duration (Fig. 11.1c)—in spite of its low corner frequency (median: 58 Hz). The answer to this apparent contradiction is found in Fig. 11.5b. This neuron may signal the presence of silent gaps larger than 2–3 ms by the reduction of its spike activity—likely in combination with at least one other neuron that signals whether sound is at all present. However, in suppressing its spike response to gappy stimuli AN4 activity disregards any specific information about the details of a stimulus' temporal structure: it represents a large range of combinations of pulse and gap durations always in the same way, i.e. by a spike rate close to zero (Fig. 11.5b). This highlights a specific filter property of this neuron, which is due to the characteristic inhibitory input to this neuron. Unfortunately, so far the source of this inhibitory input has not yet been identified.

The characterisation of neurons by means of the MTF paradigm also intended to explore if in these insects a kind of *filter bank* exists for the evaluation of amplitude modulations, similar as it has been proposed for the auditory system of mammals (Langner and Schreiner 1988; Joris et al. 2004). This relates to the question of whether the amplitude modulations of acoustic signals are processed in the time domain or possibly in the frequency domain (von Helversen and von Helversen 1998). However, in a large sample of rMTF obtained from identified neuron types we found no indication for a filter bank (Weschke and Ronacher 2008; Wohlgemuth et al. 2011). Furthermore, also behavioural experiments indicate that acoustic communication signals in grasshoppers are processed in the time domain and not in the frequency domain of amplitude modulations (von Helversen and von Helversen 1998; Schmidt et al. 2008, for a similar result in crickets see also Hennig 2009).

11.3 Discrimination of Signals in the Context of Sexual Selection

The spike train metric method and the MTF paradigm differ in one important aspect: the latter neglects the trial-to-trial variability by assessing spike rates over a longer time period (rMTF) or by using period histograms that sample spike times over many periods as basis for the tMTF evaluation. Therefore, if we want to estimate how well two similar signals can be discriminated on the basis of sensory spike trains, the knowledge of the MTF does not always help and the spike train metric is the better option.

11.3.1 Afferent Responses Allow a Good Discrimination of Similar Signals

Auditory afferents exhibit tonic responses and faithfully represent the sound envelope in their spiking pattern, up to high modulation frequencies (Fig. 11.4). Their response is rather precise and reliable, although some trial-to-trial variability is observed, depending on sound intensity and envelope characteristics (Krahe and Ronacher 1993; Machens et al. 2001; Rokem et al. 2006). Among the population of local neurons we can discern those with primary-like responses that resemble the afferents in their firing pattern and precision. Others, e.g. the BSN1 neuron show phasic-tonic to phasic firing patterns and intermediate levels of reliability and precision (Stumpner 1989; Wohlgemuth and Ronacher 2007). In contrast, the responses of ascending neurons are definitely less precise and less reliable, which seems at odds with their function to provide the brain with essential information about auditory events (Fig. 11.2 and Vogel et al. 2005; Wohlgemuth and Ronacher 2007).

An efficient signal representation should enable the discrimination of relevant signals. Returning to the discrimination of similar communication signals in the context of sexual selection, we address the question posed above: if two spike trains differ to some extent, do they still represent the same object? Whether two similar acoustic signals can be discriminated or not depends on whether the respective sensory spike trains are sufficiently different to be distinguished by down-stream neurons.

Using the songs of eight *C. biguttulus* males, we investigated how well natural calling songs of this species can be discriminated on the basis of the spike trains of auditory afferents. To remove spectral and intensity cues, the envelopes of these songs were filled with the carrier frequency spectrum of one male and presented at the same maximum intensity. In addition the length of the subunits (syllable plus pause) was equalised to remove the interindividual differences of the subunit periods as potential cues for discrimination (Machens et al. 2003). In behavioural tests *C. biguttulus* females still discriminated between these modified songs (Einhäupl et al. 2011). The same songs were used in electrophysiological recordings from auditory receptors (Machens et al. 2003) and interneurons (Wohlgemuth 2008) and it was tested how well the respective spike trains could be assigned to the different songs. For each of the eight songs one spike train was chosen as a template, and all other spike trains were assigned to that template to which they had the smallest distance according to the spike train metric. The procedure was then repeated for different template spike trains, yielding an average classification success for the song stimuli. The classification success thus indicates the proportion of the spike trains that were correctly assigned to the corresponding acoustic stimulus (for details of the procedure see Wohlgemuth and Ronacher 2007).

The result obtained with spike trains of auditory afferent neurons was remarkable: in spite of the trial-to-trial variability already single spike trains of a single afferent neuron allowed for a very good discrimination of the eight songs (classification success in the range of 90 % correct, Machens et al. 2003). This was confirmed in a study using stimuli comprising sinusoidal amplitude modulations (Wohlgemuth and Ronacher 2007, cf. Fig. 11.4a). Here the maximum classification success was somewhat lower, around 80 %, due to the fact that a third of the stimuli were amplitude modulated at high frequencies, which were beyond the corner frequencies of most neurons. The high average classification success indicates that the remaining modulation frequencies (between 10 and 167 Hz) could be perfectly discriminated on the basis of auditory afferents (Fig. 11.6). Thus, at

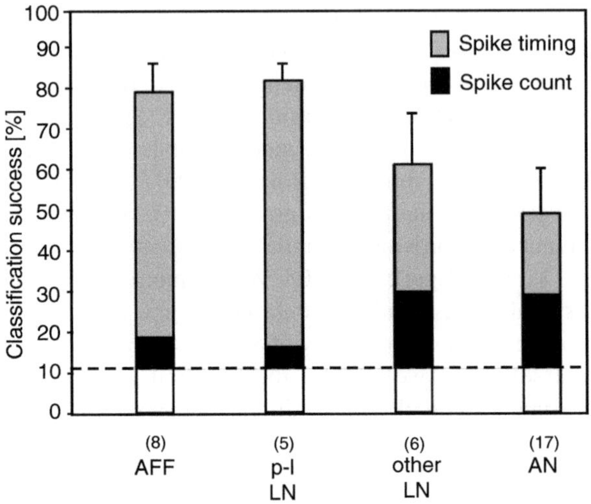

Fig. 11.6 Discrimination of amplitude modulated stimuli at different stages of the auditory pathway (*AFF* auditory afferents, *p-l LN* local neurons with primary-like responses, *other LN* other local neurons, *AN* ascending neurons). The percentage of correct classifications (discrimination success) of nine sinusoidally amplitude-modulated stimuli was determined on the basis of spike train distances evaluated with the van Rossum (2001) metric at different temporal resolutions (for details see Wohlgemuth and Ronacher 2007). Black columns show the contribution of spike count differences to the classification success, and grey columns the contribution of spike timing differences. The stippled horizontal line indicates chance level. From receptors to ascending neurons the overall classification success, and in particular the impact of spike timing information, is reduced while at the same time spike count differences become more important; adapted from Wohlgemuth and Ronacher (2007)

the input level of the system, even a single auditory afferent conveys surprisingly reliable information in its responses.

Local neurons with primary-like responses performed with the same discrimination accuracy of ~80 % correct classifications (Fig. 11.6). However, the picture changed particularly at the third processing stage, the ascending neurons: here the classification success—again based on single neurons—dropped markedly to values around 50 % correct (Fig. 11.6). This reduction was obviously due to the increased spike train variability in these neurons: both interspike interval variability as well as spike count variability increase from afferents to local interneurons and ascending interneurons (see Fig. 11.2b and Vogel et al. 2005; Wohlgemuth and Ronacher 2007; Wohlgemuth 2008).

11.3.2 The Coding Principle Changes Between Local and Ascending Neurons

The spike train metric offers the advantage to explore a continuum of different neural coding schemes. By varying the width τ of the filter function by which the spikes are replaced for the spike metric evaluation (van Rossum 2001; Machens et al. 2003) one can adjust the temporal resolution of the spike train metric, to focus, for example, on the two special cases of a *rate code*, in which only spike numbers are relevant, or on a *spike time code*, in which the temporal position of spikes as well as differences in spike count are evaluated (van Rossum 2001; Ronacher et al. 2008; Ronacher and Stange 2012). This allowed us to disentangle the respective contributions of spike count differences and spike timing to the total stimulus discrimination success (black and grey parts of columns in Fig. 11.6). For receptor neurons and local neurons with primary-like responses the discrimination depended almost exclusively on the timing of spikes. In contrast, spike count differences became more important among ascending neurons, and their contribution to the—overall lower—classification success became approximately equal to the contribution of spike timing cues (Fig. 11.6, Wohlgemuth and Ronacher 2007; Wohlgemuth 2008).

The increased relevance of spike count and the decreased contribution of spike timing among ascending neurons suggest a change of coding principles at the output stage of the thoracic processing module. Ascending neurons exhibit a higher diversity of feature selectivity and seem to be specialised to encode different features of auditory stimuli with an emphasis on rate coding. They seem to trade the "when" for "what" in their spike response wherefore the reduced spike timing precision (Fig. 11.2) becomes tolerable (Clemens et al. 2012). Prominent examples are AN4 which may signal the presence or absence of gaps (Fig. 11.5) in its spike rate, or AN12 which is thought to encode the duration of the pauses that separate the sound syllables (Fig. 11.1; Creutzig et al. 2009, 2010). The specific filter properties of other ascending neurons are less obvious, but see below. Thus, the information about a sound stimulus' envelope appears to be distributed among the set of ascending neurons. Indeed, applying a multi-neuron metric (Houghton and Sen 2008) yielded quantitative evidence for such a change of coding principles, i.e. that among ascending neurons a population code based on neurons with distinct filter properties is implemented (Clemens et al. 2011).

Unfortunately, our knowledge of the recognition centres in the brain is fragmentary. So far, no neurons with distinct species-specific filter characteristics have been described in *C. biguttulus* (in contrast to the cricket's brain: Schildberger 1994, Zorović and Hedwig 2011). However, in this context a recent modelling study on grasshoppers' song recognition has to be mentioned. Using a genetic learning algorithm (Mitchell 1998), Jan Clemens trained a set of classifiers on behavioural data obtained with a large set of artificial song stimuli. Starting from a set of 500 random solutions the feature detectors "evolved" to reproduce the behavioural data (for details see Clemens and Ronacher 2013). A characteristic of this classifier model was that the output of each feature detector was integrated over time, to yield a single value. Thus, any information about the exact temporal position of a specific song feature was omitted. The temporal integration was motivated by behavioural data with artificial songs in which syllable and pause durations were scrambled without destroying their attractiveness (von Helversen and von Helversen 1998), and fits also to the reduced spike timing precision observed in ascending neurons (Vogel et al. 2005; Wohlgemuth and Ronacher 2007; Wohlgemuth et al. 2011).

This modelling study yielded two interesting results: (i) Already two feature detectors—one excitatory, the other suppressive—sufficed to reproduce the behavioural data very well (for example the band pass tuning for pause duration, Fig. 11.1b, and additional parameters as the influence of intensity and onset accentuation), explaining 87 % of their variance (Clemens and Ronacher 2013). Inclusion of a third detector only led to a marginally further improved performance. Thus, the model gives a hint that the decision centres in the brain may rely on some cross-fibre patterns of the ascending neurons, with a focus on a spike rate code. In that scenario, the ascending neurons serve as filters that extract certain features from the sound envelopes, which then are evaluated in the brain for species-specific feature combinations. (ii) While the output of one detector showed a significant correlation with the behavioural data, the second detector, taken *per se*, did not correlate at all with the behavioural data. Only by a linear combination of both detectors the high performance of ~90 % was achieved (Clemens and Ronacher 2012). This result has important consequences when we search for neuronal correlates of behaviour: a neuron's tuning may be highly relevant for behavioural decisions, but nonetheless show no obvious correlation to the usual test stimuli. By considering each neuron in isolation we may overlook neurons with inconspicuous or variable responses that nevertheless contribute significantly to signal recognition.

11.4 The Auditory Pathway is Highly Conserved Between Grasshopper Species: Consequences for Signal Evolution

For successful communication it is essential that sender and receiver are matched to each other to at least some degree (von Helversen and von Helversen 1994). As mating success and reproduction are at the stake for grasshoppers, we could expect that their auditory pathway is specialised to process the species-specific signal patterns particularly well. The efficient coding hypothesis (Barlow 1961) predicts that coding properties of sensory neurons are optimised with respect to the relevant natural stimuli they process (Simoncelli and Olshausen 2001; Machens et al. 2005). We compared the coding properties of identified, homologous auditory neurons in two grasshoppers, *C. biguttulus* and the locust *Locusta migratoria*; two species which differ strongly in the relevance of acoustic signals for mate finding. For a most stringent, quantitative comparison we applied the van Rossum spike train metric to recordings from identified neurons. Both species were stimulated with the song of *C. biguttulus*, which is evidently not relevant for the locust. We recorded from several specimens of both species and compared the spike train similarities between homologous neurons in different specimens of the same species as well as between both species. Remarkably, the spike trains of one neuron type were not more different if we determined distances between the two species as compared to the intraspecific distances (Fig. 11.7; Neuhofer et al. 2008). This agreement suggests that the thoracic neurons and network properties were strongly

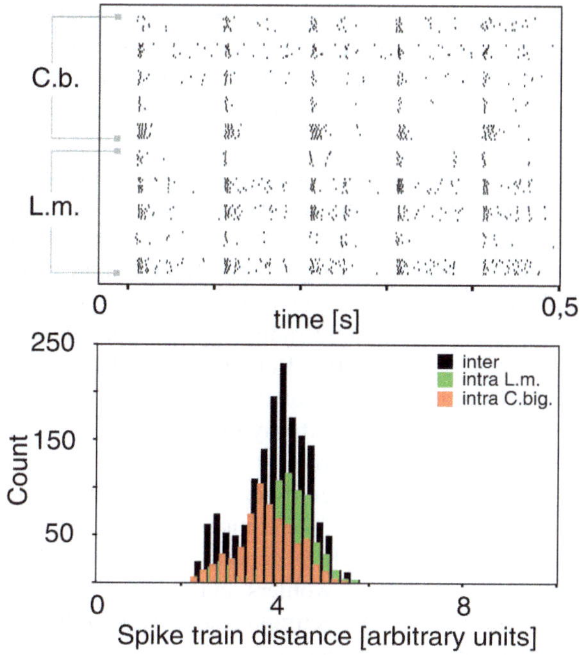

Fig. 11.7 Comparison of coding properties of homologous neurons of two grasshopper species (*Locusta migratoria*, *Chorthippus biguttulus*). *Top*: spike raster plots of responses of 5 AN12 neurons recorded in both species to repeated presentations of a natural song of *C. biguttulus*. Each point indicates the occurrence of an action potential. Distances between two spike trains were determined at high temporal resolution ($\tau = 5$ ms) for all possible combinations of spike trains with the van Rossum metric. *Bottom*: Distribution of distances between spike train combinations comparing different specimens of AN12 neurons of the locust (L.m., *green*), or between AN12 neurons of *C. biguttulus* (C. big. *orange*), or between *C. biguttulus* and *L. migratoria* (inter, *black*). Note that the *interspecific* comparison yielded no larger values than the *intraspecific* comparisons. This result was also observed with six additional neuron types. Figure from Neuhofer et al. (2008), with permission

conserved during evolution, although *C. biguttulus* and *L. migratoria* are not closely related: locusts and gomphocerine grasshoppers have been separated since about 50 Mio years (Flook and Rowell 1997). This has an interesting evolutionary implication: probably the communication signals have evolved to optimally match the properties of the sensory pathway of the receiver. The same conclusion was also drawn in relation to intensity invariance properties of grasshopper auditory neurons (Clemens et al. 2010). In contrast, according to the 'efficient coding hypothesis' (Barlow 1961; Machens et al. 2005) one would rather postulate the reverse sequence, i.e. that the processing capacities of the auditory pathway would have adapted to these highly relevant natural signals.

The postulate of the 'efficient coding hypothesis'—that sensory neurons should adjust their specific coding properties to the statistics of relevant natural stimuli—fits to the typical situation in evolution driven by natural selection, as a sensory system can not influence the properties of environmental signals. In communication systems, however, the situation may be different: the necessity for a match between signals and receiver properties entails a reciprocal coevolutionary adaptation of both parts driven by sexual selection. Thus, it depends on the respective plasticity of sender and receiver properties and on the selective pressures acting on both, whether the signals or the receiver properties may evolve more easily (Clemens et al. 2010). The neuronal hardware of the thoracic auditory pathway is obviously a highly conserved trait in grasshoppers (Neuhofer et al. 2008). In contrast, the specific communication signals of many acridid grasshopper species seem to have evolved rather recently, for example many species of the European *C. biguttulus* group underwent a rapid radiation only after the last glacial epoch (Bugrov et al. 2006; Mayer et al. 2010; Berger et al. 2010). This recent radiation may have been based on an evolution of the acoustic communication signals to match the properties of the auditory pathway of the receiver in a sensory exploitation scenario (e.g. Ryan et al. 2001; Arnqvist 2006).

The manifold of at least twenty identified types of ascending neurons in grasshoppers (Fig. 11.2a) is remarkable if we compare it with the auditory system of crickets and bush crickets, for which in general only two to three neurons have been found that ascend to the brain (Wohlers and Huber 1982; Römer et al. 1988; Stumpner and von Helversen 2001; Stumpner and Molina 2006; Triblehorn and Schul 2009). This difference between the auditory systems of Caelifera and Ensifera suggests an expansion of the parameter space which is potentially available for communication signals in grasshoppers. The capacity of a more sophisticated analysis of sound patterns may have enabled the evolution of highly complex songs in acridid grasshoppers, while crickets and bush crickets mostly produce rather simple song patterns (von Helversen 1986; Stumpner and von Helversen 1992; Gerhardt and Huber 2002; Vedenina et al. 2007; for bushcricket songs see, e.g. Heller 1988; Schul 1998; Bush et al. 2009).

Acknowledgements I thank all members of my lab who contributed to the work reported here, in particular Matthias Hennig for years of fruitful collaboration and many stimulating discussions. I also thank Daniela Neuhofer, Jana Sträter, Astrid Vogel and Sandra Wohlgemuth for providing partially unpublished data. Special thanks to Stefanie Krämer for making the figures. Berthold Hedwig and Carl Gerhardt gave valuable suggestions that helped to improve the manuscript. The American Psychological Association, the Royal Society, and Springer Science and Business Media very kindly allowed the reproduction of figures. Financial support from the German Research Council (SFB 618, GRK 837, GRK 1589/1) and the Bernstein Centre for Computational Neuroscience (Federal Ministry of Education and Research, Germany, grant 01GQ1001A) is also gratefully acknowledged.

References

Arnqvist G (2006) Sensory exploitation and sexual conflict. Phil Trans R Soc Lond B 361:375–386

Barlow HB (1961) Possible principles underlying the transformations of sensory messages. In: Rosenblith WA (ed) Sensory communication. MIT Press, Cambridge, pp 217–234

Bauer M, von Helversen O (1987) Separate localisation of sound recognizing and sound producing neural mechanisms in a grasshopper. J Comp Physiol A 165:687–695

Berger D, Chobanov DP, Mayer F (2010) Interglacial refugia and range shifts of the alpine grasshopper *Stenobothrus cotticus* (*Orthoptera*: *Acrididae*: *Gomphocerinae*). Org Divers Evol 10:123–133

Boyan GS (1999) Presynaptic contributions to response shape in an auditory neuron of the grasshopper. J Comp Physiol A 184:279–294

Brumm H, Slabbekoorn H (2005) Acoustic communication in noise. Adv Study Behav 35:151–209

Bugrov A, Novikova O, Mayorov V, Adkinson L, Blinov A (2006) Molecular phylogeny of Palaearctic genera of Gomphocerinae grasshoppers (Orthoptera, Acrididae). Syst Entomol 31:362–368

Bush SL, Beckers OM, Schul J (2009) A complex mechanism of call recognition in the katydid *Neoconocephalus affinis* (Orthoptera: Tettigoniidae). J Exp Biol 212:648–655

Clemens J, Ronacher B (2013) Feature extraction and integration underlying perceptual decision making during courtship in grasshoppers. J Neurosci 33:12136–12145

Clemens J, Weschke G, Vogel A, Ronacher B (2010) Intensity invariance properties of auditory neurons compared to the statistics of relevant natural signals in grasshoppers. J Comp Physiol A 196:285–297

Clemens J, Kutzki O, Ronacher B, Schreiber S, Wohlgemuth S (2011) Efficient transformation of an auditory population code in a small sensory system. Proc Natl Acad Sci USA 108:13812–13817

Clemens J, Wohlgemuth S, Ronacher B (2012) Nonlinear computations underlying temporal and population sparseness in the auditory system of the grasshopper. J Neurosci 32:10053–10062

Creutzig F, Wohlgemuth S, Stumpner A, Benda J, Ronacher B, Herz AVM (2009) Time-scale invariant representation of acoustic communication signals by a bursting neuron. J Neurosci 29:2575–2580

Creutzig F, Benda J, Wohlgemuth S, Stumpner A, Ronacher B, Herz AVM (2010) Timescale-invariant pattern recognition by feed-forward inhibition and parallel signal processing. Neural Comput 22:1493–1510

Einhäupl A, Stange N, Hennig RM, Ronacher B (2011) Attractiveness of grasshopper songs correlates with their robustness against noise. Behav Ecol 22:791–799

Elsner N (1974) Neuroethology of sound production in Gomphocerine grasshoppers (Orthoptera: Acrididae) 1. Song patterns and stridulatory movements. J Comp Physiol 88:67–102

Elsner N, Popov AV (1978) Neuroethology of acoustic communication. Adv Insect Physol 13:229–355

Flook PK, Rowell CHF (1997) The phylogeny of the Caelifera (Insecta, Orthoptera) as deduced from mtrRNA gene sequences. Mol Genet Evol 8:89–103

Franz A, Ronacher B (2002) Temperature dependence of temporal resolution in an insect nervous system. J Comp Physiol A 188:261–271

Gerhardt HC, Huber F (2002) Acoustic communication in insects and anurans. University of Chicago Press, Chicago

Gottsberger B, Mayer F (2007) Behavioral sterility of hybrid males in acoustically communicating grasshoppers (Acrididae, Gomphocerinae). J Comp Physiol A 193:703–714

Halex H, Kaiser W, Kalmring K (1988) Projection areas and branching patterns of the tympanal receptor cells in migratory locusts *Locusta migratoria* and *Schistocerca gregaria*. Cell Tissue Res 253:517–528

Hedwig B (1992) On the control of stridulation in the acridid grasshopper *Omocestus viridulus* L. I Interneurons involved in rhythm generation and bilateral coordination. J Comp Physiol A 171:117–128

Hedwig B (1994) A cephalothoracic command system controls stridulation in the acridid grasshopper *Omocestus viridulus* L. J Neurophysiol 72:2015–2025

Heller K-G (1988) Bioakustik der europäischen Laubheuschrecken. Josef Margraf Verlag, Weikersheim

Hennig RM (2009) Walking in Fourier's space: algorithms for the computation of periodicities in song patterns by the cricket *Gryllus bimaculatus*. J Comp Physiol A 195:971–987

Hennig RM, Franz A, Stumpner A (2004) Processing of auditory information in insects. Microsc Res Tech 63:351–374

Houghton CJ, Sen K (2008) A new multineuron spike train metric. Neural Comput 20:1495–1511

Jacobs K, Otte B, Lakes-Harlan R (1999) Tympanal receptor cells of *Schistocerca gregaria*: correlation of soma positions and dendrite attachment sites, central projections, and physiologies. J Exp Zool 283:270–285

Joris PX, Schreiner CE, Rees A (2004) Neural processing of amplitude-modulated sounds. Physiol Rev 84:541–577

Klappert K, Reinhold K (2003) Acoustic preference functions and sexual selection on the male calling song in the grasshopper *Chorthippus biguttulus*. Anim Behav 65:225–233

Klappert K, Reinhold K (2005) Local adaptation and sexual selection: a reciprocal transfer experiment with the grasshopper *Chorthippus biguttulus*. Behav Ecol Sociobiol 58:36–43

Kostarakos K, Römer H (2010) Sound transmission and directional hearing in field crickets: neurophysiological studies outdoors. J Comp Physiol A 196:669–681

Krahe R, Ronacher B (1993) Long rise times of sound pulses in grasshopper songs improve the directionality cues received by the CNS from the auditory receptors. J Comp Physiol A 173:425–434

Kriegbaum H (1989) Female choice in the grasshopper *Chorthippus biguttulus*: mating success is related to song characteristics of the male. Naturwissenschaften 76:81–82

Kriegbaum H, von Helversen O (1992) Influence of male songs on female mating behavior in the grasshopper *Chorthippus biguttulus* (Orthoptera, Acrididae). Ethology 91:248–254

Lang F (2000) Acoustic communication distances of a gomphocerine grasshopper. Bioacoustics 10:233–258

Langner G, Schreiner CE (1988) Periodicity coding in the inferior colliculus of the cat. I. Neuronal mechanisms. J Neurophysiol 60:1799–1821

Machens CK, Stemmler MB, Prinz P, Krahe R, Ronacher B, Herz AVM (2001) Representation of acoustic communication signals by insect auditory receptor neurons. J Neurosci 21:3215–3227

Machens CK, Schütze H, Franz A, Stemmler MB, Ronacher B, Herz AVM (2003) Auditory receptor neurons preserve characteristic differences between conspecific communication signals. Nature Neurosci 6:341–342

Machens CK, Gollisch T, Kolesnikova O, Herz AVM (2005) Testing the efficiency of sensory coding with optimal stimulus ensembles. Neuron 47:447–456

Marquart V (1985) Local interneurons mediating excitation and inhibition onto ascending neurons in the auditory pathway of grasshoppers. Naturwissensch 72:42–43

Mayer F, Berger D, Gottsberger B, Schulze W (2010) Non-Ecological radiations in acoustically communicating grasshoppers? In: Glaubrecht M (ed) Evolution in action. Springer, Berlin Heidelberg, pp 451–464

Meyer J, Elsner N (1996) How well are frequency sensitivities of grasshopper ears tuned to species-specific song spectra? J Exp Biol 199:1631–1642

Michelsen A (1971) The physiology of the locust ear. I. Frequency selectivity of single cells in the isolated ear. Z vergl Physiol 71:49–62

Michelsen A, Larsen ON (1983) Strategies for acoustic communication in complex environments. In: Huber F, Markl H (eds) Neuroethology and behavioural physiology. Springer, Berlin, pp 321–331

Mitchell M (1998) An introduction to genetic algorithms. MIT press, Boston

Neuhofer D, Wohlgemuth S, Stumpner A, Ronacher B (2008) Evolutionarily conserved coding properties of auditory neurons across grasshopper species. Proc R Soc Lond B 208:1965–1974

Neuhofer D, Stemmler M, Ronacher B (2011) Neuronal precision and the limits for acoustic signal recognition in a small neuronal network. J Comp Physiol A 197:251–265

Prinz P, Ronacher B (2002) Temporal modulation transfer functions in auditory receptor fibres of the locust (*Locusta migratoria* L.). J Comp Physiol A 188:577–587

Rokem A, Watzl S, Gollisch T, Stemmler MB, Herz AVM, Samengo I (2006) Spike-timing precision underlies the coding efficiency of auditory receptor neurons. J Neurophysiol 95:2541–2552

Römer H (1976) Die Informationsverarbeitung tympanaler Rezeptorelemente von *Locusta migratoria*. J Comp Physiol A 109:101–122

Römer H (2001) Ecological constraints for sound communication: from grasshoppers to elephants. In: Barth FG, Schmid A (eds) Ecology of sensing. Springer, Berlin, pp 59–77

Römer H, Marquart V (1984) Morphology and physiology of auditory interneurons in the metathoracic ganglion of the locust. J Comp Physiol A 155:249–262

Römer H, Marquart V, Hardt M (1988) Organization of a sensory neuropile in the auditory pathway of two groups of Orthoptera. J Comp Neurol 275:201–215

Ronacher B, Stange N (2012) Processing of acoustic signals in grasshoppers – a neuroethological approach towards female choice. J Physiol Paris 107:41–50

Ronacher B, Stumpner A (1988) Filtering of behaviourally relevant temporal parameters of a grasshopper's song by an auditory interneuron. J Comp Physiol A 163:517–523

Ronacher B, von Helversen D, von Helversen O (1986) Routes and stations in the processing of auditory directional information in the CNS of a grasshopper, as revealed by surgical experiments. J Comp Physiol A 158:363–374

Ronacher B, Franz A, Wohlgemuth S, Hennig RM (2004) Variability of spike trains and the processing of temporal patterns of acoustic signals – problems, constraints, and solutions. J Comp Physiol A 190:257–277

Ronacher B, Wohlgemuth S, Vogel A, Krahe R (2008) Discrimination of acoustic communication signals by grasshoppers: temporal resolution, temporal integration, and the impact of intrinsic noise. J Comp Psychol 22:252–263

Ryan MJ, Phelps SM, Rand AS (2001) How evolutionary history shapes recognition mechanisms. Trends Cogn Sci 5:143–148

Schildberger K (1994) The auditory pathway of crickets: adaptations for intraspecific acoustic communication. In: Schildberger K, Elsner N (eds) Neural basis of behavioural adaptations. G. Fischer, Stuttgart, pp 209–225

Schmidt AKD, Römer H (2011) Solutions to the cocktail party problem in insects: selective filters, spatial release from masking and gain control in tropical crickets. PLoS ONE 6:e 28593

Schmidt A, Ronacher B, Hennig RM (2008) The role of frequency, phase and time for processing amplitude modulated signals by grasshoppers. J Comp Physiol A 194:221–233

Schul J (1998) Song recognition by temporal cues in a group of closely related bushcricket species (genus Tettigonia). J Comp Physiol A 183:401–410

Simoncelli EP, Olshausen BA (2001) Natural image statistics and neural representation. Annu Rev Neurosci 24:1193–1216

Stange N, Ronacher B (2012) Grasshopper calling songs convey information about condition and health of males. J Comp Physiol A 198:309–318

Stumpner A (1988) Auditorische thorakale Interneurone von *Chorthippus biguttulus* L.: morphologische und physiologische Charakterisierung und Darstellung ihrer Filtereigenschaften für verhaltensrelevante Lautattrappen. Ph.D. thesis, Friedrich-Alexander Universität Erlangen-Nürnberg

Stumpner A (1989) Physiological variability of auditory neurons in a grasshopper. Naturwissenschaften 76:427–429

Stumpner A, Molina J (2006) Diversity of intersegmental auditory neurons in a bush cricket. J Comp Physiol A 192:1359–1376

Stumpner A, Ronacher B (1991) Auditory interneurons in the metathoracic ganglion of the grasshopper *Chorthippus biguttulus*. 1. Morphological and physiological characterization. J Exp Biol 158:391–410

Stumpner A, von Helversen O (1992) Recognition of a two-element song in the grasshopper *Chorthippus dorsatus* (Orthoptera: Gomphocerinae). J Comp Physiol A 171:405–412

Stumpner A, von Helversen O (1994) Song production and song recognition in a group of sibling grasshopper species (*Chorthippus dorsatus, C. dichrous,* and *C. loratus*: Orthoptera, Acrididae). Bioacoustics 6:1–23

Stumpner A, von Helversen D (2001) Evolution and function of auditory systems in insects. Naturwissenschaften 88:159–170

Stumpner A, Ronacher B, von Helversen O (1991) Auditory interneurons in the metathoracic ganglion of the grasshopper *Chorthippus biguttulus*. 2. Processing of temporal patterns of the song of the male. J Exp Biol 158:411–430

Triblehorn JD, Schul J (2009) Sensory-encoding differences contribute to species-specific call recognition mechanisms. J Neurophysiol 102:1348–1357

van Rossum MCW (2001) A novel spike distance. Neural Comput 13:751–763

Vedenina VY, Panyutin AK, von Helversen O (2007) The unusual inheritance pattern of the courtship songs in closely related grasshopper species of the *Chorthippus albomarginatus*-group (Orthoptera: Gomphocerinae). J Evol Biol 20:260–277

Viemeister NF, Plack CJ (1993) Time analysis. In: Yost WA, Popper AN, Fay RR (eds) Human psychophysics. Springer, Berlin, pp 116–154

Vogel A, Ronacher B (2007) Neural correlations increase between consecutive processing levels in the auditory system of locusts. J Neurophysiol 97:3376–3385

Vogel A, Hennig RM, Ronacher B (2005) Increase of neuronal response variability at higher processing levels as revealed by simultaneous recordings. J Neurophysiol 93:3548–3559

von Helversen D (1972) Gesang des Männchens und Lautschema des Weibchens bei der Feldheuschrecke *Chorthippus biguttulus* (Orthoptera, Acrididae). J Comp Physiol A 81:381–422

von Helversen O (1979) Angeborenes Erkennen akustischer Schlüsselreize. Verh Deutsch Zool Ges 1979:42–59

von Helversen O (1986) Gesang und Balz bei Feldheuschrecken der *Chorthippus albomarginatus*-Gruppe. Zool Jahrb Systematik 113:319–342

von Helversen D, von Helversen O (1975a) Verhaltensgenetische Untersuchungen am akustischen Kommunikationssystem der Feldheuschrecken (Orthoptera, Acrididae). I Der Gesang von Artbastarden zwischen *Chorthippus biguttulus* und *C. mollis*. J Comp Physiol 104:273–299

von Helversen D, von Helversen O (1975b) Verhaltensgenetische Untersuchungen am akustischen Kommunikationssystem der Feldheuschrecken (Orthoptera, Acrididae). II Das Lautschema von Artbastarden zwischen *Chorthippus biguttulus* und *C. mollis*. J Comp Physiol 104:301–323

von Helversen O, von Helversen D (1994) Forces driving coevolution of song and song recognition in grasshoppers. In: Schildberger K, Elsner N (eds) Neural basis of behavioural adaptations. G. Fischer Verlag Stuttgart, pp 253–284

von Helversen D, von Helversen O (1997) Recognition of sex in the acoustic communication of the grasshopper *Chorthippus biguttulus* (Orthoptera, Acrididae). J Comp Physiol A 180:373–386

von Helversen D, von Helversen O (1998) Acoustic pattern recognition in a grasshopper: processing in the frequency or time domain? Biol Cybern 79:467–476

Weschke G, Ronacher B (2008) Influence of sound pressure level on the processing of amplitude modulations by auditory neurons of the locust. J Comp Physiol A 194:255–265

White JA, Rubinstein JT, Kay AR (2000) Channel noise in neurons. Trends Neurosci 23:131–137

Wohlers D, Huber F (1982) Processing of sound signals by six types of neurons in the prothoracic ganglion of the cricket, *Gryllus campestris* L. J Comp Physiol A 146:161–173

Wohlgemuth S (2008) Repräsentation und Unterscheidbarkeit amplitudenmodulierter akustischer Signale im Nervensystem von Feldheuschrecken. PhD thesis Humboldt- Universität zu Berlin

Wohlgemuth S, Ronacher B (2007) Auditory discrimination of amplitude modulations based on metric distances of spike trains. J Neurophysiol 97:3082–3092

Wohlgemuth S, Vogel A, Ronacher B (2011) Encoding of amplitude modulations by auditory neurons of the locust: influence of modulation frequency, rise time, and modulation depth. J Comp Physiol A 197:61–74

Zador A (1998) Impact of synaptic unreliability on the information transmitted by spiking neurons. J Neurophysiol 79:1219–1229

Zorović M, Hedwig B (2011) Processing of species-specific auditory patterns in the cricket brain by ascending, local, and descending neurons during standing and walking. J Neurophysiol 105:2181–2194

Chapter 12
Sound Communication in *Drosophila*

Damiano Zanini, Bart Geurten, Christian Spalthoff and Martin C. Göpfert

Abstract The fruit fly *Drosophila melanogaster* communicates acoustically via courtship songs and hears with antennal ears. Research over the past decade has provided insights into the neuronal basis of *Drosophila* sound production and hearing and the functional workings of *Drosophila* ears: the neural substrate for song production has been narrowed down to subsets of Fruitless^{M} positive neurons, and the neural pathways for hearing have begun to be revealed. Mechanisms of sound transduction, adaptation, and amplification in the fruit fly's ear have been uncovered, and auditory relevant molecules have emerged from mutant screens. This chapter summarized the current state of research on *Drosophila* sound communication and hearing and discusses recent progress in the field.

12.1 Introduction

Sound communication in the fruit fly *Drosophila melanogaster* lacks some of the features that have made orthopterans and cicadas prime paradigms for insect auditory research: First, operating at milli- to centimeter distances, acoustic communication in *Drosophila* is a rather intimate affair. Second, although the fly produces loud sounds, these sounds are virtually inaudible to human ears; and third, the hearing organs of *Drosophila* lack sound-receiving tympana, whose presence is sometimes considered a defining characteristic of 'true' ears (e.g., Michelsen and Larson 1985). Here we shall see that all the above properties simply reflect the acoustic nature of *Drosophila* sound signals, and that despite these peculiarities there are many good reasons to study *Drosophila* hearing and to call its hearing organs 'ears': over the past years, parallels between *Drosophila* and vertebrate

D. Zanini · B. Geurten · C. Spalthoff · M. C. Göpfert (✉)
Department of Cellular Neurobiology, University of Göttingen, Julia-Lermontowa-Weg 3, 37077 Göttingen, Germany
e-mail: mgoepfe@gwdg.de

B. Hedwig (ed.), *Insect Hearing and Acoustic Communication*,
Animal Signals and Communication 1, DOI: 10.1007/978-3-642-40462-7_12,

hearing have been uncovered at various stages of auditory signaling and different levels of complexity, ranging from sound reception to transduction and central processing and from auditory organ performance to molecular processes and genes (e.g., Tauber and Eberl 2003; Kernan 2007; Göpfert and Robert 2008; Lu et al. 2009; Nadrowski et al. 2010). The ear of *Drosophila* affords experimental access to molecular processes in auditory sensory cells and thus provides a genetically and experimentally tractable system to investigate the integrated function of an ear. After a brief account of *Drosophila* songs and the neural basis of song production, this chapter will focus on the mechanism of sound processing in the *Drosophila* hearing organ and discuss recent advances in the study of downstream circuits in the brain. In *Drosophila*, these latter central processing mechanisms are still largely *terra incognita*, leaving room for discoveries to come. We will not cover aspects of auditory organ development in *Drosophila* and respective parallels with vertebrates, which have been reviewed by Fritzsch and Beisel (2004) and Boekhoff-Falk and Eberl (2007).

12.2 *Drosophila* Songs

Like many other drosophilid flies, *Drosophila melanogaster* communicates acoustically in the context of courtship behavior: when approaching a female, male flies typically extend—and fan—one of their wings, thereby generating a courtship song (Bennet-Clark 1971; von Schilcher 1976; Hall 1994; Greenspan and Ferveur 2000; Villella and Hall 2008) (Fig. 12.1a). This song, which is also known as the fruit fly's 'love song' (Ewing 1978), enhances the receptivity of conspecific females to copulation and stimulates conspecific males to court and sing (e.g., Crossley et al. 1995; Eberl et al. 1997). In *Drosophila melanogaster*, two main song types occur: the pulse song and the sine song (e.g., Ewing 1983) (Fig. 12.1b). The pulse song is generated by sequences of one to two rapid wing strokes that are repeated at species-specific intervals of about 35 ms. The resulting sound accordingly consists of short sound pulses that are spaced by 35 ms interpulse intervals ('IPIs') and whose dominant frequency component ranges between 150 and 200 Hz (e.g. Göpfert and Robert 2002; Riabinina et al. 2011). The sine song, in turn, arises from continuous wing fanning and consists of a distorted sinusoid with a dominant frequency around 150 Hz (e.g. Riabinina et al. 2011). Sounds at those frequencies are principally audible to our ears, but we nonetheless cannot hear these songs: The reason is that the fanning wings are rather bad sound-emitters that hardly produce any sound pressure. The sound particle velocity that is emitted by the wings, however, can be high: At a distance of ca. 5 mm, which is about the male–female distance observed during courtship, sound particle velocities have been estimated to be approximately 94 dB relative to 5×10^{-8} m/s, the latter reference value corresponding to a sound pressure level (SPL) of 0 dB (Bennet-Clark 1971).

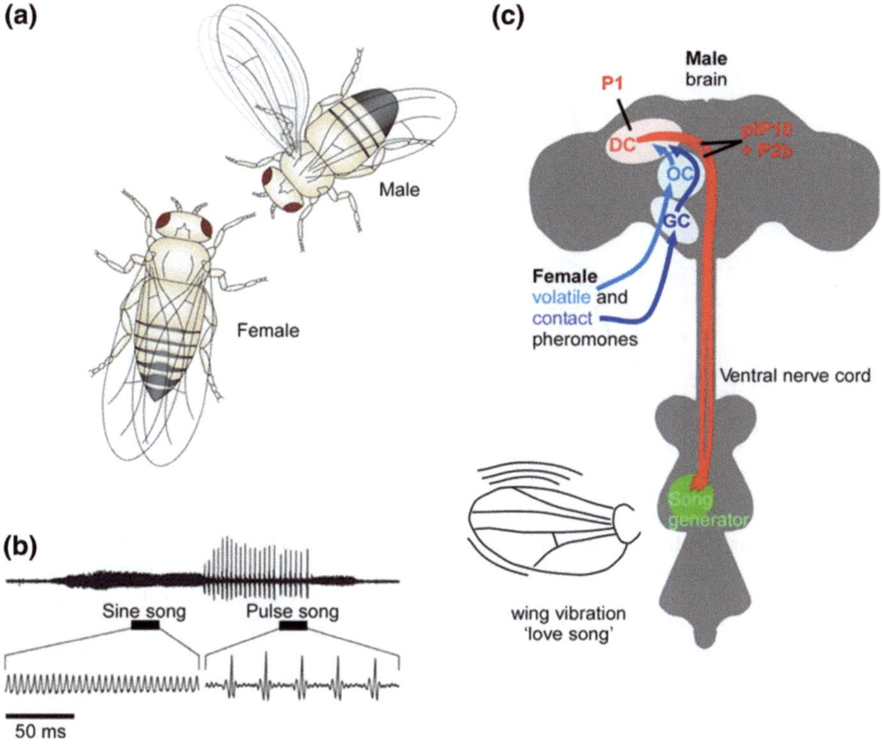

Fig. 12.1 *Drosophila* courtship song **a** Male courting a female acoustically by fanning one of its wings. Modified from Sokolowski (2001). **b** Time trace of the song depicting sine song and pulse song components, sections of which that are reproduced at an expanded time scale in the lower trace. In this example, the dominant frequency of the sine song is 157 Hz and the IPI of the pulse song averages 38 ms. Reproduced from Clyne and Miesenböck (2008). **c** Schematic circuit underlying male song production. *DC* decision center, *OC* olfactory center, *GC* Gustatory center. Modified from Benton (2011)

The neural substrate for courtship song generation by male flies has been traced down to some 20 male-specific interneurons named P1 (Kohatsu et al. 2011, von Philipsborn et al. 2011) that express the sex-specific transcription factor Fruitless^{M} (Fru^{M}) (e.g. Dickson 2008; for the requirement of Fru^{M} neurons for song production, see also Clyne and Miesenböck 2008). P1 neurons located in the lateral protocerebrum receive, and possibly integrate, visual, gustatory, and auditory input (Fig. 12.1c). Other Fru^{M} positive neurons whose activity drives song production include the plP10 and P2b neurons, which innervate the lateral protocerebrum and may act as descending command neurons that drive the production of the song (Kohatsu et al. 2011, von Philipsborn et al. 2011; for a respective commentary, see Benton 2011).

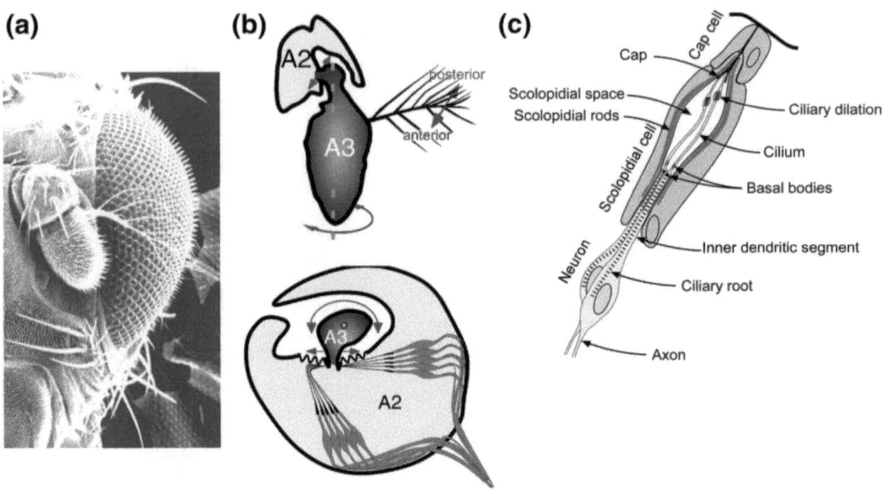

Fig. 12.2 *Drosophila* ear. **a** Frontal view of the *Drosophila* head depicting the left antenna. Reproduced from Göpfert and Robert (2002). **b** Sketch of the antenna (*top*) depicting its second (A2) and third (A3) segments and longitudinal section through A2 showing the opposing populations of JO neurons that perpendicularly connect to A3. Arrows indicate the rotational movement of A3 when the antennal arista vibrates back and forth in the presence of sound. Reproduced from Nadrowski et al. (2008). **c** Sketch of a JO sensillum with two-ciliated mechanosensory neurons. Modified from Sarpal et al. (2003)

12.3 *Drosophila* Hearing

12.3.1 Drosophila Ears

Different to hearing in parasitoid flies (see Chap. 4 by Hedwig and Robert) song detection in *Drosophila melanogaster* is mediated by the antennae (Manning 1967; Ewing 1978) which consist of four main parts each. From proximal to distal these are: the scape (i.e., the antenna's first segment), the pedicel (its second segment), the funiculus (its third segment), and the feathery arista (Fig. 12.2a, b). The arista is stiffly coupled to the side of the funiculus, and the funiculus proximally enters the pedicel where it bents into a hook (Göpfert and Robert, 2001). This funicular hook is flexibly hinged in the pedicel by joint membranes, which allow the funiculus with the arista to twist back and forth (Göpfert and Robert 2001, 2002). Mechanical measurements have confirmed that the funiculus and the arista together act as a sound receiver: when acoustically stimulated, they sympathetically twist back and forth like a rigid body about the pedicel-funiculus hinge (Göpfert and Robert 2001, 2002) (Fig. 12.2b).

Stimulus-induced movements of the antenna's sound-receiving distal part are monitored by Johnston's organ (JO), a chordotonal organ in the pedicel of the antenna (Ewing 1978; Kamikouchi et al. 2009; Yorozu et al. 2009). In *Drosophila*

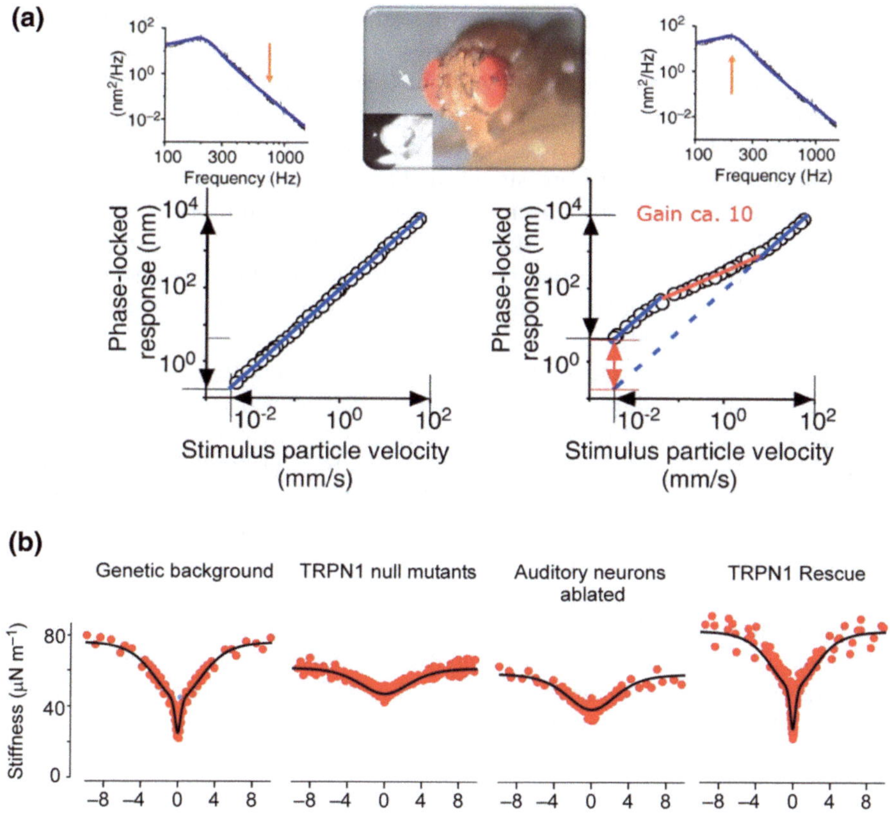

Fig. 12.3 Auditory organ function. **a** Free mechanical fluctuations (*top*) measured at the tip of the antennal arista (inset, measurement point highlighted by an *arrow*) and phase-locked displacement of the arista (*bottom*) as a function of the sound particle velocity. *Bottom left*: stimulation with pure tones at a frequencies of 800 Hz (*orange arrow, small upper left panel*) yields a linear scaling (*blue line*) of the arista displacement. *Bottom right*: stimulation with pure tones at a frequency that matches the individual best frequency of the arisa (*orange arrow, small upper right panel*) yields a compressive nonlinearity (*red line*) at intermediate sound particle velocities that enhances the arista response to faint sounds with an amplificatory gain of approximately 10. Modified from Göpfert et al. (2005). **b** Nonlinear gating compliance in the arista mechanics in genetic background controls, flies that lack the NOMPC TRPN1 channel, flies that lack auditory class AB neurons, and in TRPN1 mutants expressing a TRPN1 rescue construct in JO neurons. Reproduced from Effertz et al. (2012). The gating compliance is equally reduced if either the NOMPC TRPN1 channel or auditory JO neurons are lost, documenting that NOMPC is specifically required for transduction in these neurons

melanogaster, JO houses about 250 multicellular chordotonal sensilla that consist of two to three ciliated mechanosensory neurons and several supporting cells each (Kernan 2007) (Fig. 12.2c). The neurons are proximally suspended in the pedicel via ligament cells, and their distal mechanosensory cilia are connected to the funicular hook via extracellular caps (Kernan 2007). With respect to this latter connection, two opposing, anterior and posterior neural populations can be distinguished (Göpfert and Robert 2002): the posterior population connects to the posterior side of the hook and thus will be stretched when the antennal receiver twists forwards and compressed when it moves back. The anterior population, by contrast, connects to the anterior side of the hook and thus will be compressed when the opposing neurons are stretched. Forward movements of the antennal receiver have been shown to decrease intracellular calcium levels in the anterior population and to increase them in the posterior population, documenting alternating activities for the two opposing neuronal populations and indicating that the neurons hyperpolarize when compressed and depolarize when stretched (Kamikouchi et al. 2009).

JO neurons can be classified based on their connection with the antennal receiver and, in addition, with respect to their axonal target regions in the *Drosophila* brain: Based on these targets, at least 5 different classes of JO neurons (classes 'A–E') are distinguished (Kamikouchi et al. 2006). Calcium imaging and targeted cell ablations in conjunction with behavioral assays have shown that whereas neuron classes A and B preferentially respond to sound-induced antennal vibrations and are required for hearing, neuron classes B and C preferentially respond to sustained antennal deflections and detect gravity and wind (Kamikouchi et al. 2009; Yorozu et al. 2009). The response characteristics and functional relevance of class E neurons remains uncertain. These neurons, however, are scarce when compared to classes AB and CE, which contribute about half the fly's 500 JO neurons each (Kamikouchi et al. 2006).

12.3.2 Auditory Organ Function

Hearing starts with the conversion of sound into vibrations of the sound receiver that are transduced into electrical signals by auditory sensory cells. The conversion of sound into vibrations can be readily examined in the fruit fly's auditory system for three main reasons: first, sticking out from the body, the fly's antennal sound receiver is freely accessible to laser Doppler vibrometric measurements of its vibrations (Göpfert and Robert 2001, 2002). Second, because the entire receiver moves like a rigid body, its mechanical filter characteristics can be fully characterized by monitoring its vibrations at one measurement point (Göpfert and Robert 2001). And third, because the fruit fly's antennal receiver is very compliant, it is possible to characterize these mechanical filter characteristics without sound (Göpfert et al. 2005). In the absence of acoustic stimuli, any structure displays Brownian motion, fluctuating mechanically in response to thermal bombardment, yet only if the structure is sufficiently compliant, the amplitude of this Brownian motion will be sufficiently large to be experimentally accessed. For the fly's antennal receiver, power spectra of these free fluctuations can be well described with a simple harmonic oscillator model, yielding a resonance frequency of about 200 Hz, close to the dominant frequency of the fly's song (Göpfert et al. 2005) (Fig. 12.3a).

Whereas the antennal receivers of live flies fluctuate with a resonance around 200 Hz, they assume a resonance of about 800–900 Hz once the flies are dead (Göpfert et al. 2003). Comparably high resonance frequencies can be observed in live flies that are acoustically stimulated with loud sounds, and the resonance continuously shifts downwards to frequencies around 200 Hz when the intensity of the sound stimulus is continuously decreased (Göpfert and Robert 2003). This level-dependent shift in the receiver's mechanical tuning is associated with a compressive nonlinearity that can be seen in the receiver's responses to pure tones (Göpfert et al. 2006): for tone frequencies around 800 Hz, the receiver displacement linearly increases with the intensity of stimulation (Fig. 12.2a), but for tone frequencies around 200 Hz a nonlinear scaling is found (Fig. 12.2b). This nonlinear scaling is said to be compressive because it condenses a wide range of sound amplitudes into a narrow range of displacement amplitudes by enhancing the receiver's response to faint sound by a factor of ca. 10 (Fig. 12.2b).

The nonlinear behavior of the fly's antennal receiver arises from JO neurons, which are motile and actively pump mechanical energy into the receiver's vibration when sound is faint. The active energy contribution has been identified by comparing the receiver's fluctuations with its response to faint sounds, which revealed that the receiver is active and operates out of thermal equilibrium (Nadrowski et al. 2008). That the activity stems from JO neurons has been demonstrated using mutations that disrupt JO neurons structure and abolish amplification (Göpfert et al. 2003; Göpfert et al. 2005). Targeted cell ablations further revealed that amplification specifically arises from auditory JO neurons. Amplification remains unaffected if the population of gravity/wind-sensitive C, E neurons is ablated, but is entirely abolished by the ablation of the sound-sensitive neurons of classes A, B (Effertz et al. 2011).

In addition to reporting the motility of JO neurons, the fly's antennal receiver also reports the gating of mechanotransduction channels within these neurons. Mechanical signatures of transduction can be observed when the receiver is rapidly deflected with force steps, and manifest themselves as a fast twitch in the receiver's displacement response (Albert et al. 2007). The height of this twitch nonlinearly scales with the amplitude of forcing, reflecting an increase in the receiver's compliance at small forcing amplitudes. This nonlinear compliance (Fig. 12.3b) quantitatively conforms to the gating spring model of mechanotransduction, which assumes that transduction is mediated by force-gated ion channels that are directly activated mechanically by the pull of gating springs. According to the gating spring model, adaptation of the channels is mediated by adaptation motors that reclose the channels during maintained forcing. In the fly, the nonlinear gating compliance observed in the receiver's mechanics completely vanishes when the receiver is kept deflected, documenting that the responsible transduction channels fully adapt when the external force is maintained.

Based on the gating spring model, Nadrowski et al. (2008) have devised a simple physical model of the fly's hearing organ that links the macroscopic performance of the antennal receiver and transduction events. In this model, the receiver is represented by a simple harmonic oscillator whose spring constant reflects the linear elasticity of

JO neurons and the antennal hinge. Because JO neurons perpendicularly connect to the receiver, two opposing populations of transduction modules are symmetrically coupled to the model receiver, with each transduction module consisting of one transduction channel that occurs in series with a set of adaptation motors and one gating spring. The model was shown to comprehensively explain the performance of the fly's antennal hearing organ, quantitatively reproducing the receiver's compressive nonlinearity, its activity, its nonlinear gating compliance, properties of sound-evoked nerve potentials, and adaptation events. According to the model, active amplification in the *Drosophila* hearing organ arises from the interplay between transduction channels and adaptation motors, documenting that active amplification arises from transduction events. The model also shows that minor alterations in transduction can lead to hyper-amplification, bringing the antennal receiver into an oscillatory regime. In this state, the transduction modules will continuously oscillate the antennal receiver in the absence of sound. In the following, we will discuss the genetics basis of fly hearing and show that such self-sustained antennal oscillations can result from genetic defects.

12.3.3 Auditory Organ Genes

Several genes that are implicated in JO function have emerged from two behavioral screens. Kernan et al. (1994) screened for mutant larvae for impaired touch-responses, and Eberl et al. (1997) mutagenized flies and tested for impairments in sound-evoked behaviors. They then recorded sound-evoked potentials from the antennal nerve in all those mutants that showed impaired mechanosensory behaviors (Eberl et al. 2000), uncovering series of fly mutants with genetic alterations in JO function. Several of the responsible mutations have been molecularly characterized, and the nature and roles of the respective gene products in the fly's ear have been examined. One protein, NOMPA, turned out to be an extracellular linker protein that localizes to the extracellular caps that, surrounding the tips of the cilia of JO neurons, connect the neurons to the antennal sound receiver (Chung et al. 2001). Loss of NOMPA disconnects the neurons from the receiver, making the flies deaf (Chung et al. 2001; Göpfert and Robert 2003). Another protein, NOMPC (= TRPN1), is a transient receptor potential (TRP) channel family member (Walker et al. 2000). NOMPC localizes to the ciliary tips of JO neurons, proximal to the ciliary dilation (Lee et al. 2010; Cheng et al. 2010; Liang et al. 2011). Loss of NOMPC abolishes mechanical amplification and sensitive antennal nerve responses, but when the flies are stimulated with loud sounds, residual nerve potentials can be evoked (Eberl et al. 2000; Effertz et al. 2011). Measurements of sound-evoked intracellular calcium signals revealed that these residual sound responses are mainly contributed by gravity/wind-sensitive class C, E JO neurons, which possess NOMPC but apparently do not need it for their mechanosensory function. By contrast, auditory JO neurons of class AB largely cease responding to sound respond to sound when NOMPC is absent, explaining why active amplification is lost (Effertz et al. 2011). Examination of the nonlinear gating compliance over a wide range of stimulus amplitudes revealed that auditory and gravity/wind-sensitive JO neurons use different transducer types, and that the more sensitive auditory transducers cease to gate and are specifically disrupted mechanically if the NOMPC TRPN1 channel is lost (Effertz et al. 2012) (Fig. 12.3b). NOMPC was shown to be an essential mechanical constituent of the fly's auditory transducers, presumably forming the auditory transduction channel or a subunit thereof. NOMPC may also form the gating spring that couples forces onto these channels. In addition to a channel pore region, the NOMPC protein bears an exceptionally large number of N-terminal ankyrin residues that form –and behave like- a molecular spring (Howard and Bechstedt 2004; Sotomayor et al. 2005; Lee et al. 2006).

Apart from NOMPC, three further TRP channels have been implicated in the function of JO neurons, the vanilloid (TRPV) channels Nan and Iav and the ankyrin (TRPA) channel Pain. Nan and Iav both localize to the proximal region of the cilia of JO neurons, below the ciliary dilation (Kim et al. 2003; Gong et al. 2004). Loss of Nan leads to the loss of Iav and vice versa, suggesting that they form a heteromultimeric Nan-Iav channel (Gong et al. 2004). Without Nan-Iav, sound-evoked nerve potentials are entirely abolished, but active amplification by JO neurons persists: what happens is that the amplification becomes excessive, with amplification gains of 100 instead of normally 10 (Göpfert et al. 2006). This means that Nan-Iav negatively controls the gain of amplification, a function that has been shown to involve NOMPC (Göpfert et al. 2006). Apart from its regulatory role, Nan-Iav seems also required for propagating electrical signals from the transduction sites downwards along the cilium given the loss of sound-evoked potentials in Nan-Iav null mutants (Göpfert et al. 2006). The role of Pain in JO has not yet been studied in detail, yet it seems that this channel might be a component of the gravity/wind-transducers of JO neurons of class CE. Pain is expressed in a subset of JO neurons, and mutations in the gene encoding Pain impair gravitaxis behavior (Sun et al. 2009).

Other proteins that have been identified in JO neurons include several components of the intraflagellar transport (IFT) machinery that is essential for cilium formation and maintenance (e.g., Han et al. 2003; Lee et al. 2008), as well as a microtubule-associated protein, DCX-EMAP, that localizes to the ciliary dilation (Bechstedt et al. 2010). A chordotonal organ specific transcription factor, Fd3f, that regulates the expression of many mechanosensory relevant proteins of the cilia has recently been identified (Newton et al. 2012), and a large number of genes that are expressed in JO and implicated in JO function has emerged from a microarray screen (Senthilan et al. 2012). Among these proteins are several visual rhodopsins, which occur in the cilia of JO neurons where they seem to modulate the gating of mechanotransduction channels. How rhodopsins are activated in JO has not yet been determined, but it seems that auditory rhodopsin function does not require light (Senthilan et al. 2012).

12.4 Central Auditory Pathways and Processing

Within the brain, JO neurons of classes A–E target distinct zones in the antennal mechanosensory motor center (AMMC) (Kamikouchi et al. 2006). These axonal target regions are accordingly named AAMC zones A to E. Five types of

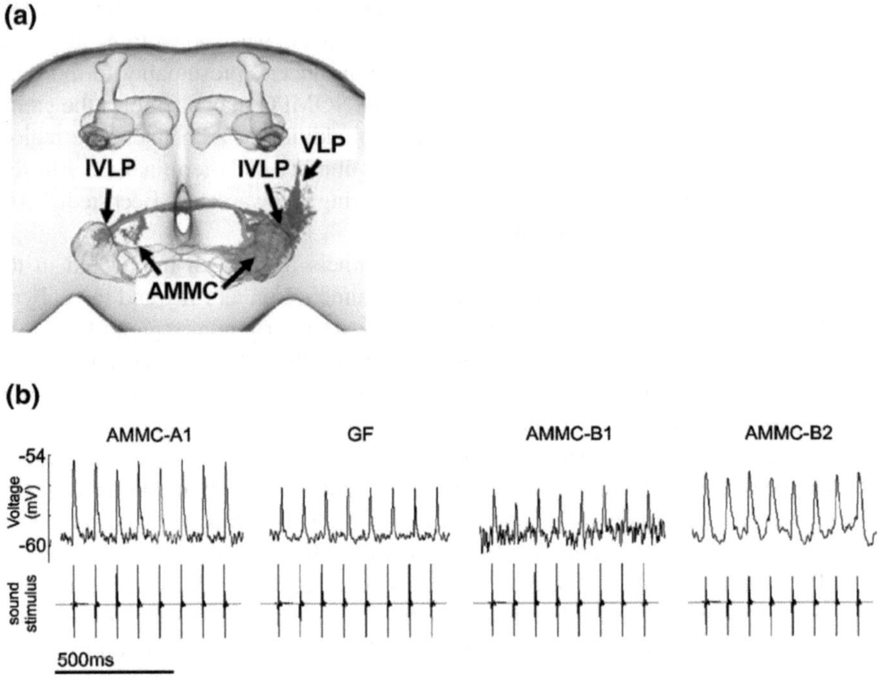

Fig. 12.4 Central auditory circuit. **a** Sketch of the *Drosophila* brain depicting the AMMC, the IVLP, and the VLP as well as some auditory interneurons. Reproduced from Lai et al. (2012). **b** Whole-cell patch clamp recordings from four second order auditory interneurons during sound stimulation. Reproduced from Tootoonian et al. (2012) with permission

second-order interneurons that arborize in the auditory AAMC zones A and B have been identified so far: (i) the giant fiber neuron that links AAMC zone A to the thoracic ganglia and the inferior ventrolateral protocerebrum (IVLP), which can be regarded as a secondary auditory center; (ii) the AMMC-A1 neuron that connects AMMC zone A and the IVLP; (iii) the AMMC-A2 neurons that extend dendrites into the AMMC zone A in both hemispheres and axonal projections into the IVLP; (iv) the AMMC-B1 neuron that links the AAMC zone B to the IVLP; and (v) the AMMC-B2 commissural neurons that interconnect the AMMC zones B of both brain hemispheres (Kamikouchi et al. 2009; Lai et al. 2012). In addition, a cluster of third-order neurons has been identified that connects the IVLP to the ventrolateral protocerebrum (VLP) (Lai et al. 2012).

Calcium imaging has confirmed that these interneurons respond to sound and, using green fluorescent protein (GFP) reconstitution across synaptic partners (GRASP) and dendritic and axonal makers, neural connectivities and the direction of the information have been deduced (Lai et al. 2012). Apparently, auditory information is first conveyed from the AMMC zones A and B to the two IVLPs and then to the ipsilateral VLP (Fig. 12.4a). VLP is composed of glomerular structures and, receiving auditory as well as visual and gustatory input, may act as a multisensory processing center (Lai et al. 2012).

Judging from in vivo whole-cell patch clamp recordings (Fig. 12.4b), AMMC-A1 and -B2 neurons are nonspiking and only generate graded potentials whereas the giant fiber neurons generate spikes (Tootoonian et al. 2012). AMMC-A1 neurons are broadly tuned and respond to courtship songs (Tootonian et al. 2012; Lai et al. 2012), whereas interneurons innervating AMMC zone B are mostly tuned narrowly to frequencies around 100 Hz. This difference in neuronal frequency tuning seems to arise at the level of JO neurons: whereas class A neurons are broadly tuned, class B neurons only respond to sounds at low frequencies around 100 Hz (Kamikouchi et al. 2009; Yorozu et al. 2009). The biological significance of this tuning difference is not yet understood, and so remain many other aspects of auditory signal processing. Measurements of antennal vibrations, for example, indicate that *Drosophila* may be endowed with directional hearing (Morley et al. 2012), but how the sound direction is encoded neutrally has hitherto not been addressed. Calcium imaging data, in turn, shows that JO neurons of classes A and B also differ in their sensitivities to sound-induced antennal vibrations (Kamikouchi et al. 2009), but what these different sensitivities are good for, likewise remains unclear. The most prominent auditory ability of *Drosophila* described so far is the distinction of conspecific songs from songs of other drosophilid species on the basis of the IPIs. The neural mechanisms underlying this IPI selectivity still awaits to be elucidated, and it seems reasonable to assume that further enticing auditory abilities will be uncovered as we learn more about acoustically evoked behaviors in the fruit fly. Menda et al. (2011) have shown that *Drosophila* can be conditioned to sounds, establishing a behavioral paradigm that might help to further probe its auditory world.

12.5 Conclusion

The study of sound communication in *Drosophila* has rapidly progressed over the past decade, providing detailed insights into the neural and genetic mechanisms of sound production and detection. Many aspects of this sound communication system, however, have not be addressed so far, leaving much room for discovery.

References

Albert JT, Nadrowski B, Göpfert MC (2007) Mechanical signatures of transducer gating in the *Drosophila* ear. Curr Biol 17:1000–1006

Bechstedt S, Albert JT, Kreil DP, Müller-Reichert T, Göpfert MC, Howard J (2010) A doublecortin containing microtubule-associated protein is implicated in mechanotransduction in *Drosophila* sensory cilia. Nat Commun 1:11

Benton R (2011) Decision making: singin' in the brain. Neuron 69:399–401

Cheng LE, Song W, Looger LL, Jan LY, Jan YN (2010) The role of the TRP channel NompC in *Drosophila* larval and adult locomotion. Neuron 67:373–380

Chung YD, Zhu J, Han Y, Kernan MJ (2001) *nompA* encodes a PNS-specific, ZP domain protein required to connect mechanosensory dendrites to sensory structures. Neuron 29:415–428

Clyne JD, Miesenböck G (2008) Sex-specific control and tuning of the pattern generator for courtship song in *Drosophila*. Cell 133:354–363

Crossley SA, Bennet-Clark HC, Evert HT (1995) Courtship song components affect male and female *Drosophila* differently. Anim Behav 50:827–839

Dickson BJ (2008) Wired for sex: the neurobiology of *Drosophila* mating decisions. Science 322:904–909

Eberl DF, Boekhoff-Falk G (2007) Development of Johnston's organ in *Drosophila*. Int J Dev Biol 51:679–687

Eberl DF, Duyk GM, Perrimon N (1997) A genetic screen for mutations that disrupt an auditory response in *Drosophila melanogaster*. Proc Natl Acad Sci USA 94:14837–14842

Eberl DF, Hardy RW, Kernan MJ (2000) Genetically similar transduction mechanisms for touch and hearing in *Drosophila*. J Neurosci 20: 5981–5988

Effertz T, Wiek R, Göpfert MC (2011) NompC TRP channel is essential for *Drosophila* sound receptor function. Curr Biol 21:592–597

Effertz T, Nadrowski B, Piepenbrock D, Albert JT, Göpfert MC (2012) Direct gating and mechanical integrity of *Drosophila* auditory transducers require TRPN1. Nat Neurosci 15:1198–1200

Ewing AW (1978) The antenna of *Drosophila* as a 'love song' receptor. Physiol. Entomol 3:33–36

Ewing AW (1983) Functional aspects of *Drosophila* courtship. Biol Rev 58:275–292

Fritzsch B, Beisel KW (2004) Keeping sensory cells and evolving neurons to connect them to the brain: molecular conservation and novelties in vertebrate ear development. Brain Behav Evol 64:182–197

Gong Z, Son W, Chung YD, Kim J, Shin DW, McClung CA, Lee Y, Lee HW, Chang DJ, Kaang BK, Cho H, Oh U, Hirsh J, Kernan MJ, Kim C (2004) Two interdependent TRPV channel subunits, inactive and Nanchung, mediate hearing in *Drosophila*. J Neurosci 24:9059–9066

Göpfert MC, Robert D (2001) Biomechanics. Turning the key on *Drosophila* audition. Nature 411:908

Göpfert MC, Robert D (2002) The mechanical basis of *Drosophila* audition. J Exp Biol 205:1199–1208

Göpfert MC, Robert D (2003) Motion generation by *Drosophila* mechanosensory neurons. Proc Natl Acad Sci USA 100:5514–5519

Göpfert MC, Robert D (2008) Active processes in insect hearing. In: Manley GA, Fay RF, Popper AN (eds) Active processes and otoacoustic emissions in hearing. Springer Handbook of Auditory Research, Springer, New, York, pp 191–209

Göpfert MC, Humphris AD, Albert JT, Robert D, Hendrich O (2005) Power gain exhibited by motile mechanosensory neurons in *Drosophila* ears. Proc Natl Acad Sci USA 102:325–330

Göpfert MC, Albert JT, Nadrowski B, Kamikouchi A (2006) Specification of auditory sensitivity by *Drosophila* TRP channels. Nat Neurosci 9:999–1000

Greenspan RJ, Ferveur JF (2000) Courtship in *Drosophila*. Annu Rev Genet 34:205–232

Hall JC (1994) The mating of a fly. Science 264:1702–1714

Han YG, Kwok BH, Kernan MJ (2003) Intraflagellar transport is required in Drosophila to differentiate sensory cilia but not sperm. Curr Biol 13:1679–1686

Howard J, Bechstedt S (2004) Hypothesis: a helix of ankyrin repeats of the NOMPC-TRP ion channel is the gating spring of mechanoreceptors. Curr Biol 14:R224–226

Kamikouchi A, Shimada T, Ito K (2006) Comprehensive classification of the auditory sensory projections in the brain of the fruit fly *Drosophila melanogaster*. J Comp Neurol 499:317–356

Kamikouchi A, Inagaki HK, Effertz T, Hendrich O, Fiala A, Göpfert MC, Ito K (2009) The neural basis of *Drosophila* gravity-sensing and hearing. Nature 458:165–171

Kernan MJ (2007) Mechanotransduction and auditory transduction in *Drosophila*. Pflügers Arch 454:703–720

Kernan M, Cowan D, Zuker C (1994) Genetic dissection of mechanosensory transduction: mechanoreception-defective mutations of *Drosophila*. Neuron 12:1195–1206

Kim J, Chung YD, Park DY, Choi S, Shin DW, Soh H, Lee HW, Son W, Yim J, Park CS, Kernan MJ, Kim C (2003) A TRPV family ion channel required for hearing in *Drosophila.* Nature 424:81–84

Kohatsu S, Koganezawa M, Yamamoto D (2011) Female contact activates male-specific interneurons that trigger stereotypic courtship behavior in *Drosophila.* Neuron 69:498–508

Lai JS-Y, Lo S-J, Dickson BJ, Chiang A-S (2012) Auditory circuit in the *Drosophila* brain. Proc Natl Acad Sci USA 109:2607–2612

Lee G, Abdi K, Jiang Y, Michaely P, Bennett V, Marszalek PE (2006) Nanospring behaviour of ankyrin repeats. Nature 440:246–249

Lee E, Sivan-Loukianova E, Eberl DF, Kernan MJ (2008) An IFT-A protein is required to delimit functionally distinct zones in mechanosensory cilia. Curr Biol 18:1899–1906

Lee J, Moon S, Cha Y, Chung YD (2010) *Drosophila* TRPN(= NOMPC) channel localizes to the distal end of mechanosensory cilia. PLoS ONE 5:e11012

Liang X, Madrid J, Saleh HS, Howard J (2011) NOMPC, a member of the TRP channel family, localizes to the tubular body and distal cilium of *Drosophila* campaniform and chordotonal receptor cells. Cytoskeleton 68:1–7

Lu Q, Senthilan PR, Effertz T, Nadrowski B, Göpfert MC (2009) Using *Drosophila* for studying fundamental processes in hearing. Integr Comp Biol 49:674–680

Manning A (1967) Antennae and sexual receptivity in *Drosophila melanogaster* females. Science 158:136–137

Menda G, Bar HY, Arthur BJ, Rivlin PK, Wyttenbach RA, Strawderman RL, Hoy RR (2011) Classical conditioning through auditory stimuli in *Drosophila*: methods and models. J Exp Biol 214:2864–2870

Michelsen A, Larsen ON (1985) Hearing and sound. In: Kerkut GA, Gilbert LI (eds) Comprehensive insect physiology, biochemistry, and pharmacology. Pergamon, New York, pp 495–556

Morley EL, Steinmann T, Casas J, Robert D (2012) Directional cues in *Drosophila melanogaster* audition: structure of acoustic flow and inter-antennal velocity differences. J Exp Biol 215:2405–2413

Nadrowski B, Albert JT, Göpfert MC (2008) Transducer-based force generation explains active process in *Drosophila* hearing. Curr Biol 18:1365–1372

Nadrowski B, Effertz T, Senthilan PR, Göpfert MC (2010) Antennal hearing in insects-new findings, new questions. Hear Res 273:7–13

Newton FG, zur Lage PI, Karak S, Moore DJ, Göpfert MC, Jarman AP (2012) Forkhead transcription factor Fd3F cooperates with Rfx to regulate a gene expression program for mechanosensory cilia specialization. Dev Cell 22:1221–1233

Riabinina O, Dai M, Duke T, Albert JT (2011) Active process mediates species-specific tuning of *Drosophila* ears. Curr Biol 21:658–664

Sarpal R, Todi SV, Sivan-Loukianova E, Shirolikar S, Subramanian N, Raff EC, Erickson JW, Ray K, Eberl DF (2003) *Drosophila* KAP interacts with the kinesin II motor subunit KLP64D to assemble chordotonal sensory cilia, but not sperm tails. Curr Biol 13:1687–1696

Senthilan PR, Piepenbrock D, Ovezmyradov G, Nadrowski B, Bechstedt S, Pauls S, Winkler M, Möbius W, Howard J, Göpfert MC (2012) Drosophila auditory organ genes and genetic hearing defects. Cell 150:1042–1054

Sokolowski MB (2001) *Drosophila*: genetics meets behaviour. Nat Rev Genet 2:879–890

Sotomayor M, Corey DP, Schulten K (2005) In search of the hair-cell gating spring elastic properties of ankyrin and cadherin repeats. Structure 13:669–682

Sun Y, Liu L, Ben-Shahar Y, Jacobs JS, Eberl DF, Welsh MJ (2009) TRPA channels distinguish gravity sensing from hearing in Johnston's organ. Proc Natl Acad Sci USA 106:13606–13611

Tauber E, Eberl DF (2003) Acoustic communication in *Drosophila.* Behav Processes 64:197–210

Tootoonian S, Coen P, Kawai R, Murthy M (2012) Neural representations of courtship song in the *Drosophila* brain. J Neurosci 32:787–798

Villella A, Hall JC (2008) Neurogenetics of courtship and mating in *Drosophila.* Adv Genet 62:67–184 von Philipsborn AC, Liu T, Yu JY, Masser C, Bidaye SS, Dickson BJ (2011) Neuronal control of *Drosophila* courtship song. Neuron 69:509–522

von Schilcher F (1976) The role of auditory stimuli in the courtship of *Drosophila melanogaster*. Anim Behav 24:18–26

Walker RG, Willingham AT, Zuker CS (2000) A Drosophila mechanosensory transduction channel. Science 287:2229–2234

Yorozu S, Wong A, Fischer BJ, Dankert H, Kernan MJ, Kamikouchi A, Ito K, Anderson DJ (2009) Distinct sensory representations of wind and near-field sound in the *Drosophila* brain. Nature 458:201–205

Index

B. Hedwig (ed.), *Insect Hearing and Acoustic Communication*,
Animal Signals and Communication 1, DOI: 10.1007/978-3-642-40462-7,

MIX
Papier aus verantwortungsvollen Quellen
Paper from responsible sources
FSC® C105338
FSC
www.fsc.org

If you have any concerns about our products,
you can contact us on
ProductSafety@springernature.com

In case Publisher is established outside the EU,
the EU authorized representative is:
Springer Nature Customer Service Center GmbH
Europaplatz 3, 69115 Heidelberg, Germany

Printed by Libri Plureos GmbH
in Hamburg, Germany